BORN IN CAMBRIDGE

BORN IN CAMBRIDGE

400 YEARS OF IDEAS AND INNOVATORS

KAREN WEINTRAUB AND MICHAEL KUCHTA

The MIT Press
Cambridge, Massachusetts
London, England

The MIT Press would like to thank the anonymous peer reviewers who provided comments on drafts of this book. The generous work of academic experts is essential for establishing the authority and quality of our publications. We acknowledge with gratitude the contributions of these otherwise uncredited readers.

This book was set in Bembo Book MT Pro by New Best-set Typesetters Ltd. Printed and bound in the United States of America.

Library of Congress Cataloging-in-Publication Data

Names: Weintraub, Karen, 1966- author. | Kuchta, Michael, author.
Title: Born in Cambridge : 400 years of ideas and innovators / Karen Weintraub, and Michael Kuchta.
Other titles: 400 years of ideas and innovators
Description: Cambridge, Massachusetts : The MIT Press, [2022] | Includes bibliographical references and index.
Identifiers: LCCN 2021013275 | ISBN 9780262046800 (hardcover)
Subjects: LCSH: Cambridge (Mass.)—Biography. | Cambridge (Mass.)—History.
Classification: LCC F74.C1 W45 2022 | DDC 974.4/4—dc23
LC record available at https://lccn.loc.gov/2021013275

10   9   8   7   6   5   4   3   2   1

publication supported by a grant from
The Community Foundation for Greater New Haven
as part of the **Urban Haven Project**

CONTENTS

We've now lived in Cambridge, Massachusetts, for more than twenty years, taking long walks and carting our daughters throughout the city's neighborhoods. Over the years, reading historical plaques and talking to our neighbors, we began keeping a mental list of "firsts" that happened here: the first college in the English colonies, the first mustering of the US Army, the first garden cemetery, the first two-way long-distance phone call, the first legal same-sex marriage. Our adopted city has also been home, at least for key points in their lives, to television chef Julia Child, civil rights leader and sociologist W. E. B. Du Bois, photography pioneer Edwin Land, and poet Robert Frost. As the list began to grow, it raised larger questions for us: how has this city of never more than 125,000 people been the scene of so much innovation and how has it managed to thrive and reinvent itself while so many other cities have struggled to stay afloat? This book represents our attempt at answers.

As we expanded our list of innovative Cambridge people and ideas, we knew we would need to make some hard choices. Cambridge's best known universities—Harvard and the Massachusetts Institute of Technology—have educated or employed hundreds of Nobel Prize–winning scientists. We realized that we couldn't write individual profiles of each, nor could we write in a compelling way about all of the prominent authors, political figures, and other celebrated Americans who have passed briefly through Cambridge. We decided that we needed to limit our list—if only to preserve our own sanity—to fewer than fifty subjects. To do this, we established some fairly tight constraints. Each topic had to be an innovation that we could credibly argue had important beginnings in the city of Cambridge. We have laid claim to the development of microwave radar, for instance, but not the microwave oven. Many of the discoveries made here have advanced specialized fields such as medicinal chemistry, theoretical physics, and mechanical engineering. As generalists, however, we wanted to focus on innovations that have resonated well beyond one particular discipline. We included the yellow highlighting marker and the Polaroid camera while leaving out sociable robots. We chose to include people, ideas, and technologies whose reputations have been solidly established rather than more recent advances that are promising but not yet fully exploited. We looked for stories that would engage readers in the complex personalities and processes of discovery that have long made Cambridge an interesting place. Many of these advances are connected to Harvard and MIT, but one of the things that surprised us was the number of important innovations that have occurred in Cambridge outside of academia.

As our list of topics grew and morphed, several general categories emerged: literature, social reform, industry and innovation, basic science, national defense, the digital world, biotechnology, and popular culture. Though the boundaries blur somewhat, we found these categories to be useful in our thinking about the unique contributions that people in Cambridge have made to the country and to the world.

What we ended up with is not a conventional history of a city. We've tried to view Cambridge through its innovators rather than through its political or community leaders. It's deliberately episodic. We haven't written much about the eighteenth century, for instance, when Cambridge had a tiny population and limited influence. Although we show that the city has been enriched throughout its history by waves of immigration, we have not focused specifically on its cultural history.

Some of our subjects were visionaries, most were exceptionally passionate, and a few were seriously misguided. Some have become household names, while others remain largely unknown. The common thread is that we found all of them interesting and worthy of exploration.

The story of Cambridge reflects the history of America. It was one of the earliest communities settled by English colonists in the 1600s. When the country was divided over slavery in the 1800s, Henry Wadsworth Longfellow was holding long conversations with abolitionists in his Cambridge living room. When World War II threatened, the city's scientists, engineers, and factory workers mobilized to develop militarily important technologies like radar and made candy for the nation's soldiers and sailors, while many families sent their sons off to fight. Major events and trends that affected the nation left fingerprints here, too.

For each of the subjects we chose to explore, we pored over biographies, journal articles, and archival records. We interviewed people who could bring the stories to life or add a modern-day perspective. And we sought out visual materials that would connect our stories to the places where they happened. We recognize that the subjects of this book don't adequately represent the true diversity of contemporary Cambridge. There are more stories to be written about creative and passionate people in this city who have made a difference in the world and make it an interesting place to live and work.

We brought to this project the perspectives of a science journalist with degrees in urban studies and political science and an architect with a longtime interest and degree in the history of technology. We like to explore connections between places and people, between the events of the past and the news of the day. After years of talking to each other about Cambridge and other cities, we've tried to draw larger lessons from our research about the nature of American urbanism, the roots of innovation, the costs and opportunities of economic success, and ways to foster the next generation of good ideas. The last chapter of the book summarizes some of those lessons.

Over the last few years, we've conscripted nearly everyone we've encountered into helping us with this project. Bumping into a friend we hadn't seen for a while, we'd mention the book and wait for their suggestions of topics we'd overlooked. Many people had suggestions; we explored them all. The Porterhouse Steak might have been invented in Cambridge's Porter Square—or a half dozen other similarly named places. Margret Rey and H. A. Rey, who invented the character Curious George, lived in Cambridge for many years but only after most of their books were written; likewise, cartoonist Al Capp moved here after he completed the work that made him famous.

The point of this is to say that we have a vast number of people to thank and that any oversights are entirely our own. We are particularly grateful to people who generously agreed to interviews, many of whose names appear in the chapters that follow. We'd like to thank Leonard Adleman; Heidelise Als; Sharon Begley and Gideon Gil of STAT; Doug Berman of *Car Talk*; Hugh Blair-Smith; Barry Bluestone of Northeastern University; Ethan Boatner; Ty Burr of the *Boston Globe*; Dan Butko at the University of Oklahoma; Jon Chase; Robin Chase; Polly Chatfield; Anna Christie of the Longfellow House–Washington's Headquarters National Historic Site; George Church; Kevin Davies; Deborah Douglas of the MIT Museum; Lance Drane; Nima and Kate Eshghi; Bob Frankston; Dan Gaier of the Missouri Valley Wrench Club; Robert Gallagher of MIT; Walter Gilbert; Owen Gingerich and Jonathan McDowell of the Harvard-Smithsonian Center for Astrophysics; author Charlotte Gordon; John Gruesser of Sam Houston State University; Margaret Hamilton; Phillip Harrington; Bree Harvey and Meg Winslow of Mount Auburn Cemetery; ink enthusiast John Hinkel; Maggie Hoffman and others at the Cambridge Historical Society; David Honn and Cynthia Hillas (children of Francis Honn); Nancy Hopkins of MIT; Christoph Irmscher of Indiana University; Mitch Kapor; Samuel Jay Keyser of MIT; Ali Khademhosseini of the Terasaki Institute; Christopher Kimball of Christopher Kimball's Milk Street; Alisha Knight of Washington College; Eric Lander and David Liu of the Broad Institute; Robert Langer of MIT; Harvey Lodish of MIT and the Whitehead Institute; Zine Magubane of Boston College; author Megan Marshall; Hope Mayo of Harvard's Houghton Library; Sheila McCullough; Victor McElheny of MIT; Matthew Meselson of Harvard; Scott Podolsky of Harvard Medical School; Mark Ptashne of Memorial Sloan Kettering Cancer Center; Gus Rancatore of Toscanini's; John Reuter; Carl Rosenberg and Benjamin Markham of Acentech Inc.; William Sahlman of Harvard Business School; William Samuelson of Boston University; Lloyd Schwartz of the University of Massachusetts Boston; Robert Socolow of Princeton University; Joshua Sparrow of Boston Children's Hospital; Jeremy Spindler of Spindler Confections; Charles Sullivan of the Cambridge Historical Commission; Lawrence Summers of Harvard University; Arun Sundararajan of New York University; Glen Urban of the MIT Sloan School of Management; J. Kim Vandiver of MIT; Kathleen Weiler of Tufts University; and Rainer Weiss of MIT.

We could not have completed this project without the repositories of historical memory that are libraries, archives, and scientific collections and the wonderful people who are their stewards. In particular, we are grateful to the Cambridge Historical Society, especially Marieke Van Damme, and for access to the Harvard University Libraries, including the Schlesinger Library at the Radcliffe Institute for Advanced Study, Houghton Library, Widener Library, the Botany Libraries at the Harvard University Herbaria, Countway Medical Library, Baker/Bloomberg Library at Harvard Business School, and Frances Loeb Library at the Graduate School of Design. Emily Gonzalez at the Cambridge Historical Commission was enormously helpful. For historical images, we relied heavily on the Cambridge Historical Commission, the Cambridge Historical Society, the MIT Museum, the National Museum of American History-Smithsonian Institution, the Rosenbach Library in Philadelphia, and Lindsay Smith Zrull at the Harvard Smithsonian Center for Astrophysics. For a variety of print materials and historic local newspapers, the Cambridge Public Library was an invaluable resource, and for maps, the Norman Leventhal Map Collection at the Boston Public Library. Ethan Boatner, Richard Howard, and Alex MacLean generously allowed us to use their photographic work. We also want to thank Charles Sullivan at the Cambridge Historical Commission and author Christoph Irmscher, who turned us into Ebay junkies in our search for visual materials to illustrate this book.

Carlynn Chapman served as an early research assistant for the project; Nell Lake was an invaluable early reader; Neta Crawford showed us to a few folks we would otherwise have missed. Nancy Sayre read and commented on portions of the book.

At the MIT Press, we are thankful for the encouragement and dedication of Amy Brand, Gita Manaktala, and especially Justin Kehoe, Rosemary Winfield, and Deborah Cantor-Adams, who encouraged and saved us—though any remaining errors are ours alone.

We would like to thank the many friends and colleagues who encouraged us and listened to our stories as this book evolved from a weird hobby into an even more time-consuming one. It was extremely helpful to have willing subjects on whom we could test-market our ideas. Of particular assistance were Holly Ambler, David Ambler, Louise Ambler, Alison Anderson, Helen Branswell, Karen Brown, Leslie Brunetta, Bob Buderi, Kathy Burge, Cathy Chute and Hull Fulweiler, Antonella Fruscione and Jeremy Drake, Lisa Girard, Carey Goldberg, Diane Gray, Dan Grossman, Devin Hahn, Katharyn Hurd, Tanya Iatridis, Susan Kahn, Noelle Karberg, Sascha Karberg and Claudia Summ, Pagan Kennedy, Nell Lake, Phil McKenna, Jane Moncreiff, Giles Moore, Kate Neptune, Jennifer Portman, Maggie Rosen, Ellen Ruppel Shell, Steve Smith, Joshua Sokel, Doug Starr, Katherine Stewart, Farah Stockman, William Doss Suter, Mark Verkennis, Elizabeth Weise, Sylvia Welsh, Chris Woodside, and Rachel Zimmerman.

Finally, we would like to thank our family members—especially our children. We have monopolized several years' worth of family dinners and neighborhood walks with conversations about this book, and we realize they will never get those hours back.

INTRODUCTION

The three-story red brick building at 700 Main Street in Cambridge, Massachusetts, isn't eye-catching. A utilitarian structure built in the 1880s, its regular rows of green double-hung windows and brickwork arches suggest a workaday industrial use. But it would be hard to find a better spot in Cambridge to illustrate the city's ability to foster innovation and reinvention.

Over the course of two centuries, the 700 Main Street site, roughly halfway between Central and Kendall squares, has hosted a remarkable array of people and ideas, some of which have had a significant impact on American society and culture. The center-aisle rail car was invented here, as was the modern adjustable pipe wrench

700 Main Street, Cambridge. Photograph by the authors, 2020.

and the Polaroid camera. Alexander Graham Bell placed the first two-way long-distance telephone call to a factory complex that once occupied the lot. In recent years, the location has housed an incubator for young biotechnology companies. Over time, each tenant has taken advantage of the site's inherent advantages: cheap land close to Boston, good access to a highly skilled workforce, a robust building that can carry the weight of heavy machinery, and a community of leaders, workers, and neighbors willing to contribute to and take a chance on something new.

Although this site at Main and Osborne streets has a particularly rich history, it's not hard to find evidence of other "firsts" around the city, across a range of disciplines and centuries: the first printed book in the English colonies, the first smallpox vaccinations in America, the first email message using the @ symbol. Cambridge has also been home to notable trailblazers: poet Henry Wadsworth Longfellow, pioneering journalist Margaret Fuller, sociologist and civil rights activist W. E. B. Du Bois, Zipcar cofounder Robin Chase, and violinist Yo-Yo Ma. Even George Washington slept here (for nine months). For a city that has never had a population of more than 125,000 to spawn this many innovative people and ideas, something unusual must be going on.

Part of the answer clearly rests with the city's academic institutions. Harvard University, founded in Cambridge in 1636, and the Massachusetts Institute of Technology, which moved here from Boston in 1916, have attracted creative, ambitious people from across the globe and provided them the resources to be successful. Around 1900, Ohio native Wallace Sabine invented the field of architectural acoustics in his Harvard physics lab—later using it to design Boston's Symphony Hall. Half a century later, another Ohio native, Harvard chemist Louis Fieser, invented napalm, the military incendiary weapon, in a building nearby. MIT linguist Noam Chomsky published "The Responsibility of Intellectuals," an essay that helped make him a leading voice in the anti–Vietnam War movement.

But many of Cambridge "firsts" have nothing to do with the city's academic institutions. The first modern sewing machine, the Fig Newton, Junior Mints, the Hi-Liter marker, and the nineteenth-century ice industry are all linked to Cambridge's history as a diversified manufacturing center.

In a city as compact as Cambridge, just seven square miles, people bump into each other. Biologists get to know chefs who befriend entrepreneurs who live next door to poets. As a young woman in the 1850s, feminist and domestic reform advocate Melusina Fay Peirce studied with natural scientist Louis Agassiz and recalled visits to her family home by Henry Wadsworth Longfellow. In the 1970s, Julia Child bought the cookware for her public television show from Cambridge architect Ben Thompson, who designed the Faneuil Hall Marketplace in Boston—in the process, inventing a new kind of urban retail venue. Chance interactions between innovators and early adopters seem to happen more frequently in a densely settled walkable place like Cambridge than they do elsewhere.

Many American cities the size of Cambridge have been unable to sustain themselves after losing their initial geographic advantage or founding industry. In the first half of the twentieth century, Cambridge might have been mistaken for Camden, New Jersey, or Gary, Indiana—burgeoning industrial centers on the periphery of major American cities. Like Camden, which sits across a river from Philadelphia, and Gary, twenty-five miles from Chicago, Cambridge hemorrhaged industrial jobs and population after World War II. But by 2020, Cambridge had nearly reclaimed its 1950 population peak, while Gary's population remained at less than half its postwar high, and Camden had lost more than 40 percent of its residents. Today, Cambridge remains one of the safest cities of its size, with a robust real estate market and well-resourced public schools.

From a colonial college town to an industrial city to a scientific research center and biotechnology powerhouse, Cambridge has managed to reinvent itself several times. The settlement in 1630 of what was first called Newtowne was established on the north banks of the Charles River to serve as the capital of the newly founded Massachusetts Bay Colony. However, Boston—closer to the sea and with a rapidly growing population—was found to be a more suitable location for the colony's government.

Newtowne's first reinvention came just six years later. As a consolation prize of sorts for losing its capital status, the town was designated by the colonial legislature as the seat of a college and renamed Cambridge. In 1638, the fledgling Newe College was renamed for a deceased Puritan minister, John Harvard. Centered on what became Harvard Square, the community of Old Cambridge grew slowly over the next two centuries. By 1820, the town had fewer than 3,500 residents.

Cambridge's next reinvention began around that time, as large-scale industrial activity took hold. Glassmaking started in East Cambridge in 1818. In the 1840s, Nathaniel Wyeth launched a brick-making business in the clay pits of North Cambridge, the beginnings of an industry that would come to employ thousands of newly arrived immigrants over the next hundred years. In the 1880s and 1890s, a local entrepreneur named Charles Davenport built a seawall along the Charles River, with the intention of creating public parks along the water's edge. By dredging the river bottom and filling in tidal areas, Davenport and his Charles River Embankment Company created many acres of inexpensive new land in Cambridge, property that was eventually occupied with factories and the MIT campus.

By 1900, Cambridge had become the Boston region's primary manufacturing center, a position it held for half a century. But after World War II, industries began leaving for the suburbs and for states with lower-cost labor, or they foundered because of fundamental market changes.

The next reinvention had already begun. At Harvard, Charles William Eliot had turned the college into a major research institution modeled after Europe's most scientifically advanced universities. Eliot, who served as Harvard's president from 1869 to 1909, reshaped its academic curriculum, dropping the Greek and Latin requirement while widening the range of courses so each member of the all-male student body could pursue his natural talents. Eliot was influential in creating standardized entrance examinations, emphasizing academic merit over social status as a basis for admission. A chemist by training, he vastly expanded Harvard's programs in the sciences, promoting basic research and the creation of new knowledge rather than the recitation of ancient truths or vocational training. Under Eliot, Harvard became the preeminent research university in America, attracting students and faculty from around the country and, increasingly, the world. Nearby, Radcliffe College began providing collegiate instruction to women in 1879, often employing the same professors who taught the young men a few blocks away. The two colleges began an extended courtship that resulted in a 1999 merger.

More seeds of reinvention were planted in 1916, when the Massachusetts Institute of Technology relocated to Cambridge from Boston's Back Bay. In the 1930s, president Karl Compton and his deputy Vannevar Bush reshaped MIT from a vocational training school to a research-focused science and engineering institute.

During World War II, both MIT and Harvard mobilized in the nation's defense, turning out microwave radar technology, high-resolution aerial photography, safer and better-fitting headphones for airplane pilots, and dozens of other mission-critical innovations. Money poured in from the federal government for basic and applied research, leading to an explosion of digital technologies in the decades following the war. Early computers like the Mark I at Harvard and Whirlwind at MIT led to network communications protocols and the first emails and ultimately to applications like Facebook and internet acceleration.

Though funding levels vary from year to year, federal research money continues to fuel Cambridge's innovation-based economy. In the last half century, Cambridge has been at the center of the revolution in molecular biology—the growing understanding of life at a cellular and genetic level—and its application to medicine and pharmaceutical research. The biopharma industry has been responsible for the city's most recent reinvention.

It is the great variety of innovation in Cambridge—generated by academic research labs as well as tinkerers, entrepreneurs, social critics, writers, and problem-solvers—that inspired this book.

In the last chapter, we outline eight reasons we think Cambridge has been so successful at reinventing itself. Some of these are specific to Cambridge, such as being close to Boston's financial, political, and population center. But others can be adapted and used by communities elsewhere in the world: a focus on education and an educated community; urban density; enlightened leadership; and diversity of population, industries, and funding streams (in Cambridge's case, including government grants, private philanthropy, and venture capital).

We think that moral intentions matter, too, and we've focused particularly on stories of Cambridge people who have innovated in the service of a larger purpose. When, in 1632, Anne Bradstreet wrote the first poem in British North America from her home in what is today Harvard Square, she was expressing her religious faith through poetry. In the 1840s, Cambridge-born writer Margaret Fuller sought to expand the educational and career opportunities available to women in American society as she simultaneously worked to support her family. In the 1940s, former MIT administrator Vannevar Bush aimed to defeat fascism by harnessing the scientific might of the country's academic institutions for the purpose of national defense. In the 1960s and 1970s, Harvard biologist Matthew Meselson sought to protect the world from the threat of inexpensive biological and chemical weapons, pressing US presidents to ratify new international treaties.

A NOTE ON HOW TO READ THIS BOOK

This book is arranged thematically rather than strictly chronologically, divided into eight main parts, each around a common theme. Chapter 1, The Literary-Industrial Complex, profiles some of the prose writers, poets, printers, publishers, and book sellers who have inhabited the city from its earliest days. Chapter 2, Social Reform in the People's Republic, describes people who set out to change the world around them, working to expand the rights of women, African Americans, and same-sex couples. Chapter 3, Industrial Cambridge, includes some of the unique products invented or manufactured here. Chapter 4, Origins of a Research Enterprise, covers scientific ideas—and a few entire fields of study—that trace their origins to Cambridge, particularly in the nineteenth century. Chapter 5, Cambridge Goes to War, reviews the role that the city's academic and manufacturing sectors played in defending the nation during World War II and some of the longer-term implications of that work. Chapter 6, Digital Cambridge, focuses on advances in computer technology and the internet. Chapter 7, Harnessing Biology, profiles people who have contributed to our understanding of the genetic basis of all life and have explored its implications for human health. Chapter 8, Our Fair City in Popular Culture, describes local residents who have created—or themselves become—cultural icons, including the game of football, television chef Julia Child, and the public radio show *Car Talk*. The last chapter, Re: Invention, explores how the city has managed to nurture so many talented people and how its urban form and diverse population have influenced its creative output.

Whether you read this book cover to cover or one story at a time, we hope you will come away with the same sense of fascination we experienced in researching and writing it.

THE LITERARY-INDUSTRIAL COMPLEX

In 2013, the New York auction house Sotheby's advertised the sale of a book it compared to the Gutenberg Bible and Shakespeare's first folio.[1] *The Whole Booke of Psalmes, Faithfully Translated into English Metre*, commonly called the *Bay Psalm Book*, was the first book ever printed in British North America. Sotheby's was selling one of only eleven surviving copies and the first on the market since 1947. After two and a half minutes of bidding, it sold for $14.2 million—nearly $3 million above the previous record for a book at auction. Translated into English from the Biblical Book of Psalms, the poems in it were intended to be sung as hymns in church. The book was one of an estimated 1,700 that had been printed nearly four centuries earlier in what is now Harvard Square.

Since its founding in 1630, the city of Cambridge, Massachusetts, has housed nearly every element of the publishing industry: a robust community of popular and scholarly authors, editors, publishers, printers, book sellers, libraries, readers—even an ink producer. If it involved words on a page, there was a good chance you'd find it in Cambridge.

Poetry and Prose

Published writers have been a presence in Cambridge since its start, when **Anne Bradstreet,** Colonial America's first major poet, became one of its first residents. Oliver Wendell Holmes Sr., the physician and poet, was born in Cambridge in 1809, graduated from Harvard College in 1829, and in later life wrote regularly for the *Atlantic Monthly* magazine—a name he chose.

Henry Wadsworth Longfellow became one of nineteenth-century America's most famous people based on the popularity of his poems, which told stories of heroes both real and imagined, from the midnight ride of Paul Revere to the mythical song of Hiawatha. The Brattle Street home where he lived for forty-five years is now a National Historic Site.

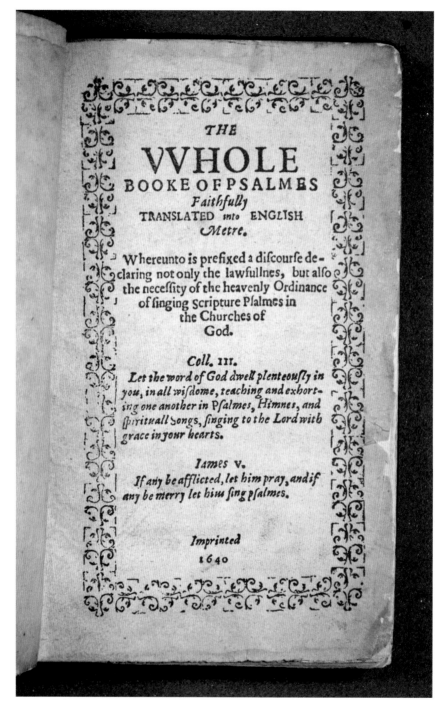

THE

VVHOLE
BOOKE OF PSALMES
Faithfully
TRANSLATED *into* ENGLISH
Metre.

Whereunto is prefixed a difcourfe de-
claring not only the lawfullnes, but alfo
the neceffity of the heavenly Ordinance
of finging Scripture Pfalmes in
the Churches of
God.

Coll. III.
Let the word of God dwell plenteoufly in
you, in all wifdome, teaching and exhort-
ing one another in Pfalmes, Himnes, and
fpirituall Songs, finging to the Lord with
grace in your hearts.

Iames v.
If any be afflicted, let him pray, and if
any be merry let him fing pfalmes.

Imprinted
1640

Title page of *The Whole Booke of Psalmes, Faithfully Translated into English Metre*, 1640.
Source: Courtesy of Rosenbach Library, Philadelphia, PA.

Pauline Elizabeth Hopkins, one of the most prominent Black authors at the turn of the twentieth century, wrote and edited articles for the influential *Colored American Magazine* while living on Clifton Street.

In the twentieth century, an astonishing number of critically acclaimed poets lived in Cambridge, including T. S. Eliot, e. e. cummings, Robert Frost, Elizabeth Bishop, and Adrienne Rich. The poet **Robert Lowell**, though he never lived permanently in Cambridge, formed the nucleus of a group of influential American poets who frequented Harvard Square's cafes and bookshops in the 1960s and 1970s.

Margret Rey and H. A. Rey, creators of the Curious George children's books, which they wrote from 1939 to 1966, moved to Harvard Square in 1963.[2] Prolific mystery novelist Robert B. Parker made Cambridge his home until his death in 2010. Lois Lowry, author of the Newbery Medal–winning *The Giver* and other young adult novels, lived in a book-filled house in Cambridge for many years.[3] Experimental psychologist Steven Pinker, the author of ten books and numerous articles in the *New York Times* and *The Atlantic*, moved to Cambridge in 1976 and has bounced between MIT and Harvard for more than four decades.[4]

Printers and Publishers

In the 1630s, dissenters from the established English church were prohibited from printing religious materials in their home country, and some Puritan congregations resorted to printing in the Netherlands.[5] Establishing a press in the Massachusetts Bay Colony would provide the Puritans a base from which to smuggle religious pamphlets and books to other colonies and back to England, free from censorship and potentially at substantial profit.

Joseph Glover, a Puritan minister, set sail from England in 1638 with his wife and five children, bringing a printing press, some secondhand metal type, and book paper, intending to start a publishing operation in the New World. Stephen Daye and his two sons came as Glover's indentured servants, obligated to work until the cost of their sea passage was repaid. The Reverend Glover died during the voyage, leaving responsibility for the press to his widow, Elizabeth. She settled in Cambridge with her five children and set up the Dayes—and likely the press—in a house on Crooked Lane, now Holyoke Street, between Massachusetts Avenue and Mount Auburn Street.[6] As the owner of the printing press, Elizabeth Glover became the first publisher in the English colonies.

In his first year operating the press, Stephen Daye printed pamphlets and an almanac of astronomical data for the Bay Colony's farmers, fishermen, and sailors. In 1640, Daye undertook the printing of the *Bay Psalm Book*, which contained "new translations made by Puritan clergymen in Massachusetts," according to Hope Mayo, a lecturer at Harvard's Houghton Library, which houses most of the university's rare book collections, including a copy of the *Bay Psalm Book*.[7] "They evidently didn't want to use translations of the psalms that the Anglicans used," she says. Daye was hardly an expert printer. The text is lighter in some places than others; some pages are out of order, with a handwritten note to see the previous leaf first. Reprinted many times, the book remained in use for well over a century.

The outbreak of the English Civil War in 1642 eliminated the tight restrictions on printing in England, and the Daye press lost its international market. Instead, the only printing press in Massachusetts for nearly fifty years was used to print the Bay Colony's laws, annual almanacs, and Harvard students' theses.[8] A plaque on Dunster Street in Harvard Square commemorates its operation here.

A variety of small printing houses continued to operate in Cambridge in the seventeenth and eighteenth centuries. Publishing began to take place at an industrial scale here in the mid-nineteenth century. In 1852, Henry Oscar Houghton founded the Riverside Press to print books for Boston's publishing houses. His main client, Little, Brown and Company, had its printing plant, warehouse, and shipping office in

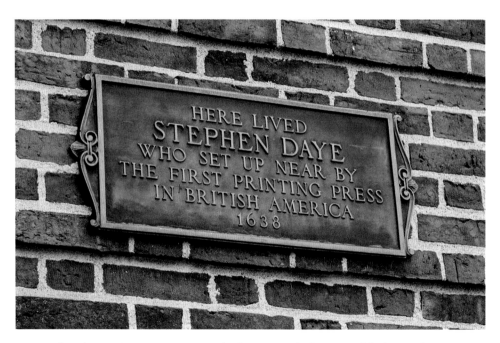

Historical marker on Dunster Street in Cambridge, noting the location of the home of printer Stephen Daye. *Source:* Photo by the authors.

Dresden Press, at one time believed to be the press of Stephen Daye, ca. 1908. *Source:* Vermont Historical Society, Prints and Photographs Division, Library of Congress, Washington, DC.

Cambridge. Houghton took on a partner, George Mifflin, in 1872, and the renamed Houghton, Mifflin and Company enlarged its printing plant on the banks of the Charles River. By 1886, the company employed more than six hundred workers locally. Houghton's insistence on readable type, elegant page and cover designs, and sturdy bindings elevated the craft of bookmaking and earned the company a reputation for high-quality literary books, eventually including the works of Henry Wadsworth Longfellow, Ralph Waldo Emerson, Nathaniel Hawthorne, Henry David Thoreau, Harriet Beecher Stowe, and Henry James. The Riverside Press operated

Interior view of the Riverside Press, Cambridge, ca. 1905. *Source:* The Riverside Press Collection #P001, Cambridge Historical Commission.

in Cambridge until 1972, when the building was razed and the site turned into a public park.

Booksellers and Libraries

In 1849, **John Bartlett**, a voracious reader, took over a bookshop in Harvard Square, which served the local academic community. He published the first edition of his *Familiar Quotations* in 1855, and the book soon became a standard reference work for librarians, term paper writers, and speech makers. Bartlett's book, now in its eighteenth edition, is still regularly updated and in print.

Despite the decline of independent bookstores across the country in the wake of big chains and e-commerce, there were still at least a dozen retail bookshops in Cambridge as of early 2021. Until it closed its Harvard Square location in 2017, Schoenhof's Foreign Books claimed to be the oldest seller of foreign language books in America, dating to 1856. The Grolier Poetry Book Shop, occupying a small storefront on Plympton Street, was for many years one of the only poetry-specific bookstores in the country.

Grolier Poetry Bookshop, Harvard Square, Cambridge, 2015. *Source:* Photo by the authors.

And then there are the libraries. Harvard University's library system, with more than sixty individual libraries, is the third largest in the United States by volumes held and has been labeled the largest privately owned library system in the world.[9] The city's main public library normally bustles seven days a week from opening to closing. Add the twenty-two Little Free Libraries[10] (as of this writing) that have popped up on sidewalks and in front yards, and it's clear that literary life still thrives in Cambridge.

ANNE BRADSTREET: COLONIAL POET

In 1997, Harvard University named one of the gates that encircle Harvard Yard in honor of Anne Bradstreet, the first published poet in British North America. Bradstreet's words are inscribed on one of the brick piers that bracket the gate: "I came into this country where I found a new World and new manners at which my heart rose." A visitor today might assume that Bradstreet, who arrived in New England at age eighteen in 1630, was excited to enter this uncharted continent, with its boundless resources and untapped potential. But to Bradstreet, the words "at which my heart rose" meant not that she was thrilled but that she was nauseated.[11] The prospect of being in the New World terrified her.

Bradstreet Gate, Harvard Yard, Cambridge, 2020. *Source:* Photo by the authors.

Inscription on Bradstreet Gate, Harvard Yard, Cambridge, 2020. *Source:* Photo by the authors.

Bradstreet biographer Charlotte Gordon chuckles as she tells the story. It's ironic, she says, that a woman whose work was underrated for centuries had been misunderstood again.[12] Gordon, an English professor at Endicott College in Beverly, Massachusetts, considers Bradstreet a true genius and a rebel who cloaked her revolutionary ideas in her Puritan religion.[13]

Bradstreet was among the first Americans of European descent to settle in Cambridge, then called Newtowne, where she lived with her husband a few years after arriving from England. She wrote the earliest of her published poems in 1632, while living approximately where the Harvard Coop now stands. Her poetry was first published in London in 1650.

Bradstreet wasn't quite twenty when she fell ill and thought she might die.[14] After she recovered, she wrote a poem, "Upon a Fit of Sickness," about the experience of cheating death yet again. (She had survived smallpox shortly before marrying.) Her brush with death and the subsequent birth of her first child, a healthy boy she named Samuel, left Bradstreet feeling that she had finally done right in God's eyes.[15]

Though no portraits of Bradstreet exist from her own time, she was depicted in later centuries as a pious bonnet-wearing housewife and mother. But she was not simply

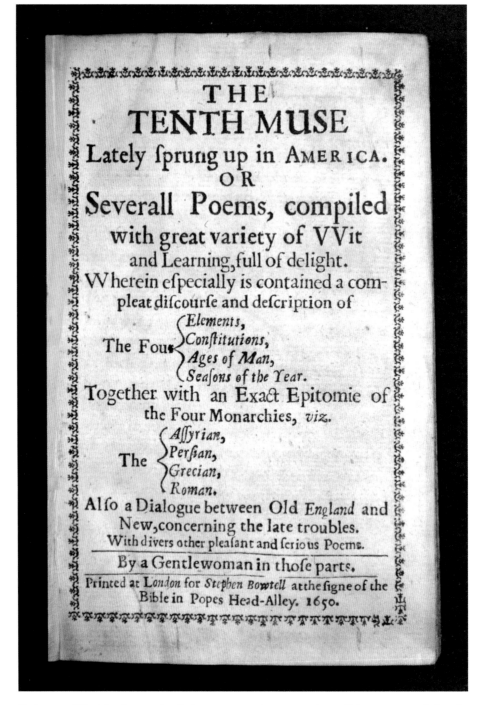

THE
TENTH MUSE
Lately sprung up in AMERICA.
OR
Severall Poems, compiled
with great variety of VVit
and Learning, full of delight.
Wherein especially is contained a com-
pleat discourse and description of

The Four
- Elements,
- Constitutions,
- Ages of Man,
- Seasons of the Year.

Together with an Exact Epitomie of
the Four Monarchies, viz.

The
- Assyrian,
- Persian,
- Grecian,
- Roman.

Also a Dialogue between Old England and
New, concerning the late troubles.
With divers other pleasant and serious Poems.

By a Gentlewoman in those parts.

Printed at London for Stephen Bowtell at the signe of the
Bible in Popes Head-Alley. 1650.

Title page, "The Tenth Muse Lately Sprung up in America, or Severall Poems, Compiled with Great Variety of Wit and Learning, Full of Delight," "By a Gentlewoman in Those Parts," 1650. *Source:* Rosenbach Library, Philadelphia, PA.

that, Gordon says; she was also a literary radical. Bradstreet wrote in a plainer and more direct language than was common in England in the mid-seventeenth century. Bradstreet's poetry has had a "lasting impact on our cultural identity," Gordon says, suggesting that the work of later American poets like Walt Whitman and Robert Frost directly descends from Bradstreet's plain-spoken style. "She's the first person to give real voice to what it meant to be a new English person," Gordon says.[16]

Poetry was a logical choice for Bradstreet's literary expression.[17] The main forms of public communication in her time were poems and sermons, but the latter would have been considered off-limits to Bradstreet, Gordon says. There were no female ministers in seventeenth-century England, and it would have seemed presumptuous for a woman to write sermons.[18] The first American novel wouldn't be written for more than another century and a half: William Hill Brown's *The Power of Sympathy; or, The Power of Nature*, published in Boston in 1789.[19] So what was left to her was poetry.

Bradstreet was born in Northamptonshire, England,[20] and raised in Sepringham Manor on the country estate of the Earl of Lincoln. Her father, Thomas Dudley, was the earl's steward. A bright but sickly child, Bradstreet was her father's favorite, and he took care to educate her well. She could read and—more unusual for her time—write by age ten, and her father schooled her in history, literature, theology, and the Scriptures. Dudley told the young Anne that "writing and Christianity went hand in hand, that the poet's job was not simply to invent a line of pentameter but to consider how best to serve God while reading and composing."[21] Her mother's teachings—on how to be a good wife and mother—took place on the hours-long weekly rides to and from church.

Lincolnshire, which included the town of Boston, was then the center of England's Puritan community and a "cauldron of dissent," according to Gordon.[22] Their minister, John Cotton, was, along with Bradstreet's father and a handful of others, among the leading Puritan thinkers of the time. Cotton believed in simple ceremonies and simple language and, like all Puritans, opposed the Catholic Pope. Cotton preached a utopian vision of a "new" England untainted by the influence of Rome.[23] He intended his ideas to be metaphors and wanted to reform England from within. But Dudley apparently took Cotton literally, imagining—and beginning to talk with his peers—about starting fresh in the New World. (Cotton would later join the Dudleys and Bradstreets in Massachusetts and become the leading minister in the new colony.)[24]

Persecuted by the British government, many Puritans eventually lost hope that they could ever reform the system from inside. The only path to redemption was to start over. Dudley was among the leaders of this group and so devoted to the cause that he committed not only to moving to America but to taking his wife and five children with him.[25] Anne, eighteen, the second-oldest, had been married nearly two years earlier to Simon Bradstreet, a promising young man who had been Anne's father's assistant throughout much of her childhood. The two, apparently, were very much in love.

Though Dudley and Simon Bradstreet were committed to restarting civilization in unchartered territory, Anne was not.[26] The Bradstreets arrived in New England aboard a ship called the *Arbella* in 1630. At the time, more than half the Europeans to have ever reached the continent had died of starvation. Her father's close friend and sometime rival, John Winthrop, would soon become the first governor of the Massachusetts Bay Colony. But when the Dudleys and Bradstreets first landed near Salem, Massachusetts, there was little more than a cluster of dugout huts and some half-starved men. The advance party of Puritans who came in 1629 had failed to build houses. The *Arbella*'s great patroness, Lady Arbella, did not survive the summer.[27]

After two months in the near-failed colony, Dudley moved his family south to Charlestown, now part of Boston. In 1630, having survived a winter in which about two hundred settlers died and an equal number returned to England, Dudley relocated his family again, a few miles away to a place he named Newtowne.[28] For himself, Dudley selected a double lot near the Charles River and built the town's first home. Anne and her husband chose a site further upland, backing onto what would become the new town's graveyard. Today, the Harvard Coop bookstore sits on the approximate site of the Bradstreet plot.

In addition to founding Newtowne, Thomas Dudley served four terms as the Massachusetts colony's governor and was a magistrate in the colonial courts. He was also a founder of Harvard College. His daughter did not remain long in Newtowne. Anne and her husband, Simon, departed for Ipswich, Massachusetts, in 1634 and later to the area now known as North Andover, where she lived to the age of sixty. Simon was heavily involved in the administration of the Bay Colony, and after Anne's death, he served as the colony's governor.

We misunderstand Bradstreet, Gordon says, if we imagine that as a housewife and mother of eight, she spent all her days trapped at home. In the mid-seventeenth century, particularly with a husband who traveled nearly nonstop, Bradstreet would have been out in public regularly, tending her garden, trading for goods, and caring for livestock. "To be a housewife in the seventeenth century meant chasing your wild cow through brambles and coming across Indians," Gordon says. Bradstreet lacked the public titles of her father and husband but would still have had high standing in the community.

In what little spare time she could claim, Bradstreet penned her poetry. Twentieth-century poet Adrienne Rich wrote that Bradstreet's poems were about the humble

Historical marker on Dunster Street in Cambridge noting the site of the home of town founder Thomas Dudley, father of poet Anne Bradstreet. *Source:* Photo by the authors.

aspects of daily life. "No more Ages of Man, no more Assyrian monarchs; but poems in response to the simple events in a woman's life: a fit of sickness; her son's departure for England; the arrival of letters from her absent husband; the burning of their Andover house."[29] Bradstreet was also incredibly brave. To survive in the new colony, "heroism was a necessity of life," Rich wrote in an introduction to a book of Bradstreet's work. "To find room in that life for any mental activity which did not directly serve certain spiritual ends, was an act of great self-assertion and vitality," Rich said. "To have written poems, the first good poems in America, while rearing eight children, lying frequently sick, keeping house at the edge of wilderness, was to have managed a poet's range and extension within confines as severe as any American poet has confronted."

Bradstreet wanted her work to be published and read, Gordon says.[30] Her brother-in-law, the Reverend John Woodbridge, published her poems anonymously in London in 1650 under the title of *The Tenth Muse Lately Sprung Up in America*. The author was indicated only as "A Gentlewoman in those parts."

Bradstreet's poems were popular in England during and just after her lifetime but faded into obscurity shortly thereafter.[31] By the late nineteenth century, her work was not much more than a historical curiosity. Harvard professor Charles Eliot Norton, who wrote the introduction to an 1897 edition of her poems, damned them with faint praise: "Now and then a single verse shows a true, if slight, capacity for poetic expression."[32]

The restoration of Bradstreet's reputation began in 1912 when Conrad Aiken included her works in his anthology of American poetry.[33] John Berryman's poem "Homage to Mistress Bradstreet," written in 1959, shows that her work had an influence on modern poets. The feminist commitment to uncovering "lost" women in history gave a boost to her reputation in the 1960s and 1970s; Rich published her foreword to *The Works of Anne Bradstreet* in 1967. Bradstreet's poems continue to be featured in American anthologies of poetry and are frequently read at wedding ceremonies.

Though Gordon published her biography of Bradstreet in 2005, her eyes still shine when she speaks of the poet's accomplishments. "She was an ambitious, forward-thinking, really complicated woman," Gordon says. "We should know who she is, and we should understand who she is."

Upon a Fit of Sickness
by Anne Bradstreet, 1632[34]
Twice ten year old not fully told
since nature gave me breath,

My race is run, my thread is spun,
lo, here is fatal death.
All men must die, and so must I;
this cannot be revoked.
For Adam's sake this word God spake
when he so high provoked.
Yet live I shall, this life's but small,
in place of highest bliss,
Where I shall have all I can crave,
no life is like to this.
For what's this life but care and strife
since first we came from the womb?
Our strength doth waste, our time doth haste,
and then we go to th' tomb.
O bubble blast, how long can'st last?
that always art a breaking,
No sooner blown, but dead and gone,
ev'n as a word that's speaking.
O whilst I live this grace me give,
I doing good may be,
Then death's arrest I shall count best,
because it's Thy decree;
Bestow much cost there's nothing lost,
to make salvation sure,
O great's the gain, though got with pain,
comes by profession pure.
The race is run, the field is won,
the victory's mine I see;
Forever known, thou envious foe,
the foil belongs to thee.

HENRY WADSWORTH LONGFELLOW'S AMERICAN MYTHOLOGY

Henry Wadsworth Longfellow, who lived in Cambridge for fifty years, was the most popular American poet of the nineteenth century and one of the country's most famous people, period. Like Walt Disney a century later, he created enduring cultural symbols for a nation that craved them and was seen as a connection to a simpler, more wholesome past. "In the early years of the Civil War, to his American middle-class readership, Longfellow meant, broadly speaking, not only social prestige, intellectual distinction, and moral authority, but the three together filtered through a radiant personal aura inseparable from his poems," a recent biography notes.[35] As another biographer remarks, "we have no real equivalent for the kind of fame that was Longfellow's during his lifetime."[36]

Longfellow was born in Portland, Maine, in 1807, and studied at Bowdoin College. After three years of travel in Europe—primarily France, Spain, Italy, and Germany, where he rapidly acquired foreign language skills—he accepted a teaching post at Bowdoin in 1829. In 1834, he was offered a professorship at Harvard, teaching Spanish and French. Prior to taking the job, he returned to study in Europe with his wife, Mary Potter. Tragically, she died on the journey after a late-term miscarriage. Longfellow had her body shipped back to America to be buried in Mount Auburn Cemetery. During this second visit to Europe, Longfellow studied the language and literature of Germany, Denmark, Sweden, Finland, and the Netherlands. There was probably no other American at the time with his command of European languages.

In December 1836, Longfellow arrived in Cambridge to teach modern languages at Harvard. The next summer, he began lodging at the Craigie House, the Georgian Revival mansion on Brattle Street where he would live the rest of his life. According to legend, Longfellow asked the widow Elizabeth Craigie, owner of the stately home, if he could board there. He was impressed by its association with George Washington, who had lived in the house for nine months in 1775 after taking command of the Continental Army. Craigie initially mistook Longfellow for an undergraduate rather than a professor. He convinced her that he was a published author by pointing to a copy of his 1835 book, *Outre-mer, A Pilgrimage beyond the Sea*, which sat on her bookshelf.[37] The house itself became an icon, a model for Colonial Revival homes built across America in the late nineteenth and early twentieth centuries, and is now a National Park Service–operated historic site. In 1843, Longfellow married Boston textile heiress Frances "Fanny" Appleton, whose father purchased Craigie House for them as a wedding present. The couple had six children together.

In Cambridge, Longfellow's career and reputation began to flourish. Here he had access to fellow authors as well as publishers and booksellers. His first collection of poems, *Voices of the Night*, was printed in 1839 by a Cambridge publisher named John Owen.[38] His poems "The Village Blacksmith" and "The Wreck of the Hesperus" were printed in book form in 1841, also by Owen, and became widely popular. In 1847, he

Longfellow House–Washington's Headquarters National Historic Site, 2020. *Source:* Photo by the authors.

published *Evangeline*, considered to be the first important long poem in American literature.[39] The epic poem tells the fictional story of a French-speaking Canadian woman who is forcibly separated from her new husband by the British during the French and Indian War (1754–1763). After wandering America for decades searching for him, Evangeline is reunited with her beloved Gabriel, shortly before his death in a Philadelphia hospital. The sentimental poem was perhaps influenced by Longfellow's own heartbreak at the death of his young wife a decade earlier.

In 1854, Longfellow's income from his published poetry, together with his wife's substantial inheritance, allowed him to retire from teaching at age forty-seven. In the years before the Civil War, he was virtually the only professional author and full-time poet in America.[40]

Henry Wadsworth Longfellow in his library at Craigie House, Cambridge. *Source:* Courtesy of
Prints and Photographs Division, Library of Congress, Washington, DC.

Longfellow's commercial and critical triumphs continued: in 1855, he published
The Song of Hiawatha; a thousand copies were sold within a few months.[41] By 1857, not
yet at the peak of his popularity, Longfellow had written eleven books of poetry that
had already sold more than 300,000 copies.[42] *The Courtship of Miles Standish*, published
in 1858, also sold well both in the United States and England.

In 1861, Longfellow's beloved second wife, Fanny, died in a tragic accident. She was
using melted wax to seal an envelope filled with her child's hair, creating a keepsake.[43]
Her clothes caught fire, though it remains unclear exactly how this happened. Longfellow rushed to quench the flames, burning his hands and face, but his efforts were futile,
and Fanny died the next day of her injuries.[44] A month later, he wrote to Fanny's sister:
"How I am alive after what my eyes have seen, I know not. I am at least patient, if not
resigned; and thank God hourly—as I have from the beginning—for the beautiful life
we led together, and that I loved her more and more to the end." He wore a beard for
the remainder of his life as a way to cover the scars on his face.[45]

Early in 1861, as southern states were planning to secede from the Union, Longfellow published "Paul Revere's Ride" in the *Atlantic Monthly*. He wrote the tale of America's early struggles for nationhood as a lesson for his own times, says Anna Christie, a ranger at the Longfellow House–Washington's Headquarters National Historic Site. She compares "Paul Revere's Ride" to the musical *Hamilton*, first performed in 2015.[46] "If you call yourself American, this is a story you share about the founding of this country, but really it's about our current struggles," she says. Longfellow saw himself as spreading the culture that the nation's founders worked so hard to achieve, Christie says.

The poem begins:

Listen, my children, and you shall hear
Of the midnight ride of Paul Revere.
On the eighteenth of April, in Seventy-Five:
Hardly a man is now alive
Who remembers that famous day and year.

He said to his friend, "If the British march
By land or sea from the town to-night,
Hang a lantern aloft in the belfry-arch
Of the North-Church-tower, as a signal-light,—
One if by land, and two if by sea;
And I on the opposite shore will be,
Ready to ride and spread the alarm
Through every Middlesex village and farm,
For the country-folk to be up and to arm."

Though never a vocal abolitionist himself, Longfellow shared many of the antislavery views of his close friend, US Senator Charles Sumner.[47] At night, Longfellow biographer Christoph Irmscher says, activists would turn up at Longfellow's door looking for help, and he often gave money to a Boston woman, Susan Hillard, who was hiding escaped enslaved people in her attic.[48] "We now know for sure that Longfellow was part of that," Irmscher says.[49]

The gregarious Longfellow assembled a network of widely known and well-regarded friends, a virtual *Who's Who* of nineteenth-century New England, including the scientist Louis Agassiz, poets James Russell Lowell and Oliver Wendell Holmes Sr., novelist Nathaniel Hawthorne, and philosopher Ralph Waldo Emerson.

Longfellow's study at the Longfellow House–Washington's Headquarters National Historic Site, Cambridge, 2018. *Source:* Photo by the authors.

But Longfellow was often an outlier among his social peers in the 1860s. "The Civil War was an enormous problem for him," Irmscher says, "because he hated everything about slavery but also was a pacifist."[50] He once spent several hours in his study meeting with a woman who had been the last enslaved person to live at Craigie House. "It's remarkable that he met with her at all," Irmscher says. "There's absolutely no record of what transpired between the two. . . . [But] that to me shows that within the confines of who he was, what his social position was, what he felt his civic obligation was, he was trying to somehow negotiate these things."

As a teacher, Longfellow introduced new approaches to language education and comparative literature, suggesting that his students learn languages from native speakers, for instance. He compared works by different authors, like Dante and Goethe. At the time, "no one was doing this," Irmscher says. He was in many ways what today might be called a multiculturalist, exploring and valuing the literary traditions of non-Anglo societies.

Many of the poems that Longfellow crafted at his home in Cambridge became part of the popular mythology of the United States, recited by American schoolchildren for decades. "Paul Revere's Ride" turned the poem's namesake from a minor player in the American Revolution to one of its best-known heroes. His epic 1855 poem *The Song of Hiawatha* chronicled the adventures of a fictional American Indian, based very loosely on Native American folktales.

Longfellow's fame as a poet extended across America and reached England. In 1868, Longfellow was invited to Buckingham Palace to visit Queen Victoria. "Everyone from the servants through [the Queen] knew him," Christie says.[51] "To him that was the greatest compliment. He was loved by the average person."

Longfellow died in 1882. He was buried in Mount Auburn Cemetery between his two beloved wives. His poetry continued to be popular after his death. It is no accident that Longfellow—the descendant of *Mayflower* passengers—and his Colonial-era home were transformed into enduring symbols of America. Biographers Irmscher and Robert Arbour suggest that Longfellow's publishers at the Boston firm of Houghton, Mifflin positioned his work as part of an American literary canon at a time when the cultural legacy of America's early Anglo-Saxon settlers seemed at risk.[52] A resurgence of American patriotism, beginning in the centennial year of 1876 and continuing into the next century, was in part a reaction to massive waves of immigration, industrialization, and urbanization that were transforming the nation's character.

Though he was still widely read, Longfellow's reputation suffered in the twentieth century among the literary avant-garde. Unlike the work of modern poets such as T. S. Eliot and Ezra Pound, Longfellow's poetry is sentimental rather than ironic, idealized rather than personal, conventionally metrical rather than written in free verse. His status as an American icon diminished as the country became more ethnically diverse and less wedded to moral certainties. For many, Longfellow became an embarrassing and preferably forgotten relic of New England's white Anglo-Saxon Protestant cultural dominance.

But Longfellow was not just a symbol to be either revered or dismissed; he was a complex human being with a compelling personal history, Irmscher says. The biographer first became interested in the poet after accidentally retrieving a collection of Longfellow's papers at Harvard's Houghton Library while researching another topic.[53] The box contained a series of drawings by Longfellow about a character he had invented, Mr. Peter Piper. "I had not expected anything like it," Irmscher says. But he was charmed. "I discovered he'd done these things for his kids, and he'd hidden them

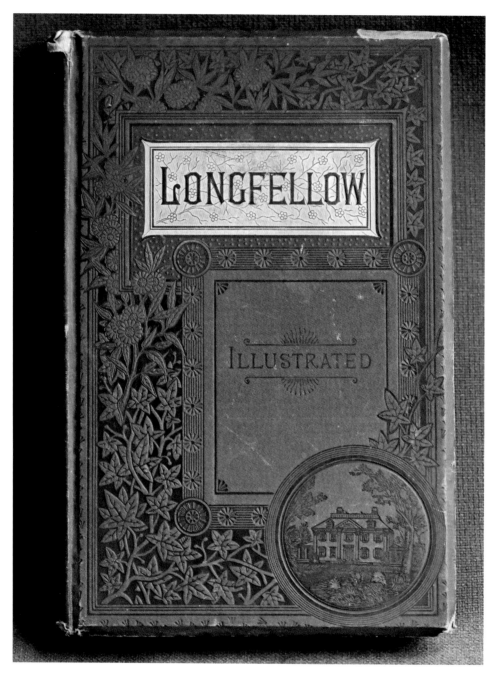

Cover of *The Poetical Works of Henry Wadsworth Longfellow*, published by Houghton, Mifflin and Company in Boston and printed by the Riverside Press in Cambridge in 1887. *Source:* Photo by the authors.

in drawers to let them find them. That didn't fit the image I had for [Longfellow]," says Irmscher, a professor of English at Indiana University. In addition to being a poet, Longfellow was a single father for most of his daughters' lives.

Today, there are public school buildings named for Longfellow in Cambridge and in Portland, Maine, which have direct ties to the poet's life, but also in places as far flung as Dallas and San Antonio, Texas; Indianapolis, Indiana; Milwaukee, Wisconsin; Bozeman, Montana; and Pasadena, California.

JOHN BARTLETT'S QUOTATIONS

Long before there was Google, there was John Bartlett.

Beginning around 1850, in the days before public libraries were common, people would come by Bartlett's Harvard Square bookshop to research, learn, and argue. Bartlett had a prodigious memory and had read and retained most of the publications on his bookshelves. He could recite Shakespeare's soliloquies and Lincoln's recent addresses. Professors would cross the street from Harvard College to ask his advice or for details from his handwritten collection of "passages, phrases and proverbs."[54]

"Here every day gathered the literati of Cambridge," a 1905 obituary in the *Cambridge Chronicle* read.[55] "Here every day occurred friendly bouts among the young and old scholars, and Bartlett was often called on to settle disputes. After a while, whenever anybody was in doubt about the origin or accuracy of a familiar quotation, he was told to 'ask John Bartlett.'" Bartlett hated not being able to answer those questions and began keeping records of quotations. Eventually, at his friends' urging, he published these as *Familiar Quotations* in 1855, a modest 267-page book.[56] Still in print today, its updated 1,500-page eighteenth edition was published in 2012, more than 150 years after the original.[57]

Author Justin Kaplan, who edited the sixteenth edition of *Bartlett's Familiar Quotations* in 1992 and the seventeenth in 2002, noted that "John Bartlett lived to see his name become as generic for quotations as Noah Webster's for definitions."[58]

Bartlett was born in 1820 in Plymouth, Massachusetts, into an established New England family. One ancestor on his father's side arrived in England along with William the Conqueror.[59] Another predecessor, born in 1603 in England, emigrated to the Plymouth Colony two decades later.[60] Bartlett's mother's family had a dramatic early introduction to the New World, as he once described it. "My maternal ancestor, Antony Thacher, arrived at Ipswich, Mass., in July 1636, and a month later was wrecked, in a violent storm, on an island near Salem, which bears his name to this day. He and his wife were the only survivors."

Portrait of John Bartlett from *Familiar Quotations*, 4th edition, 1863. *Source:* Collection of the authors.

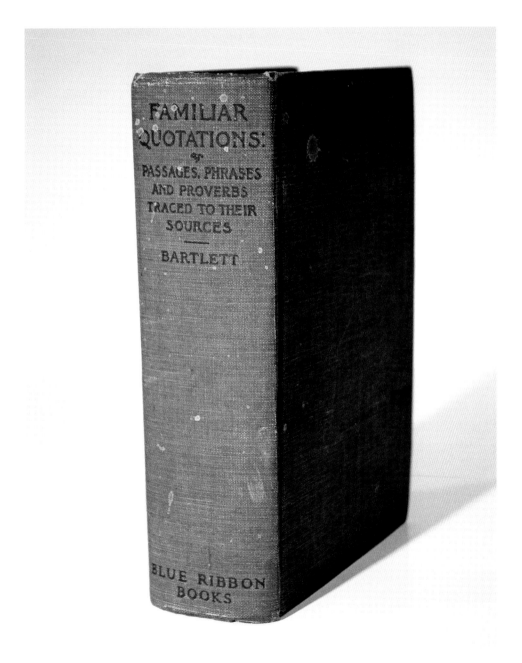

John Bartlett's *Familiar Quotations*, 10th edition, 1919. *Source:* Collection of the authors.

Bartlett inherited his Puritan ancestors' work ethic, their "sturdy independence," and what today we would call grit.[61] He also possessed a bottomless well of cheer and a keen sense of humor, according to his friend, lawyer Joseph Willard.

As a boy, Bartlett learned his letters early and read voraciously, consuming by age twelve most of the children's and adult literature available to him, including *The Pilgrim's Progress*, *Arabian Nights*, and *The Last of the Mohicans*.[62]

In 1836, at age sixteen, Bartlett moved to Cambridge, where he went to work as a bookbinder at the University Book Store, located at the corner of Massachusetts Avenue and Holyoke Street.[63] Bartlett later wrote that he found himself "amid a world of books in wandering images lost. Without a guide, philosopher, or friend. I plunged in, driving through the sea of books like a vessel without pilot or rudder."[64] Despite the tasks that were loaded on his young shoulders, Bartlett always found time to read. Sixty-four years later, in 1900, he compiled a list of all the books he'd read since turning seventeen. The eighty-five handwritten pages catalog just under seventeen hundred books, or about twenty-seven each year. In a memorial address given shortly after Bartlett's death in 1905, his friend Willard suggested that Bartlett's wide-ranging reading habits were driven by his desire to compensate for the lack of a college education.[65]

Bartlett took over the University Book Store in 1849.[66] It became a regular stopping place for Harvard students and professors, including some of its best-known: poets Henry Wadsworth Longfellow and James Russell Lowell, naturalist Louis Agassiz, and mathematician Charles Sanders Peirce.[67] An avid fly fisherman, Bartlett once gave a gift to James Russell Lowell, his Brattle Street neighbor, which Lowell acknowledged with a poem, "To Mr. John Bartlett—Who Had Sent Me a Seven-Pound Trout."[68] Bartlett regularly sent Lowell fish "for years thereafter."[69]

People were drawn into Bartlett's inner circle by his personality, strong tastes in literature, and support for intellectual pursuits. Bartlett was generous to the young students who couldn't always afford the books they wanted, and he was simply fun to be around, a friend and neighbor, Woodward Emery, later recalled. "He once told me he never felt despondent or downhearted. Certainly, cheerfulness was a pronounced characteristic."[70]

In 1859, Bartlett sold the bookstore. He served as a volunteer in the Civil War, after which he joined the Boston publishing house Little, Brown and Company as an editor, becoming a partner in the firm in 1865.[71]

The fourth edition of *Familiar Quotations*, published in 1863, topped five hundred pages. The book's popularity encouraged Bartlett to keep it updated. "The favor shown to former editions has encouraged the compiler of this collection to go on with the work and make it more worthy," he wrote, at the same time apologizing for its length. "It has been thought better to incur the risk of erring on the side of fulness."[72]

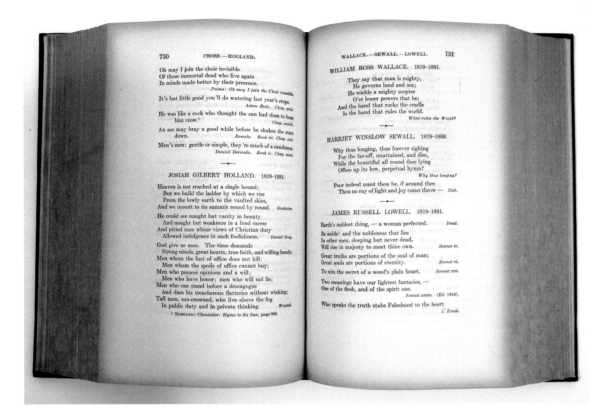

Quotations of several nineteenth-century authors, including Cambridge poet James Russell Lowell, in John Bartlett, editor, *Familiar Quotations*, 10th edition, 1919. *Source:* Collection of the authors.

The eighth edition in 1883 had 912 pages, and the ninth, the last edition that Bartlett himself edited, ran to nearly 1,200.[73] Bartlett added context along with length, so in these later editions, readers could learn the origins of each quote, not just its most famous usage.

Bartlett was devoted to his wife of more than fifty years, Hannah Staniford Willard, the daughter of a Harvard professor and granddaughter of a Harvard president.[74] She helped him keep track of citations and footnotes. "Her care and cheerful aid in arranging the thousands of slips of quotations made that great work possible for him to accomplish," his friend Emery told the Cambridge Historical Society.[75] The Bartletts never had children, but that, Emery said, "seemed to make them all the more dependent on each other—a loyal, happy, and united marriage, with an old-time halo of sacred love encircling it."

For his service to the academic community, Harvard College awarded Bartlett an honorary master of arts degree in 1871. In 1892, Bartlett was named a fellow of the American Academy of Arts and Sciences, a Cambridge-based learned society. The Academy recognized Bartlett for "the career of one who was not a trained scholar in any of the sections among which our membership is divided, nor a practitioner in any of the learned professions, but a man who in the course of an active and successful mercantile life of fifty-two years found time to devote to literature in every one of them."[76]

PAULINE ELIZABETH HOPKINS AND THE *COLORED AMERICAN MAGAZINE*

At the turn of the twentieth century, Pauline Elizabeth Hopkins was the most prominent Black female writer in America. A novelist, journalist, playwright, and editor, from 1900 to 1904 Hopkins served as editor of the *Colored American Magazine*, a Boston-based monthly and America's first Black literary periodical. In 1900, she published her best-known work of fiction, *Contending Forces: A Romance Illustrative of Negro Life North and South*, an epic historical novel that traces the experiences of several generations of an African American family. Three more novels, serialized in the *Colored American Magazine*, followed. In her fiction and nonfiction essays and in her work as a magazine editor, Hopkins tackled issues of race relations, social history, and the role of women. She also was the first African American—man or woman—to write a play that was staged with actors rather than simply read aloud, says John Gruesser, a senior research scholar at Sam Houston State University in Texas.[77] "As a literary artist, she's just amazing," he says.

For much of her adult life, Hopkins lived in Cambridge, first in a home she purchased on Clifton Street in North Cambridge and later in rented rooms in Cambridgeport.

"She was really fundamental to the continued development of Black literature after the Civil War," says Alisha Knight, a Hopkins scholar and associate professor at Washington College in Maryland.[78] Hopkins's importance, Knight says, came from her role at the *Colored American Magazine*, as well as the sheer volume of work she produced and the number of genres she contributed to.

The *Colored American* had one of the largest readerships of any literary publication oriented to African Americans until the NAACP's *Crisis* magazine supplanted it.[79] Hopkins served variously as its literary editor, editor-in-chief, and frequent contributor of published pieces. The eighty-eight-page March 1903 issue, for example, featured three pieces by Hopkins, one written under a pseudonym. The magazine included advertisements from businesses as far afield as Richmond, Virginia, and East Saint Louis, Illinois.[80] "The reach of that publication across the country was really ahead of its time," Knight says. "That's why she's so important to me as a scholar and teacher."

(See page 218.)

Portrait of Pauline Elizabeth Hopkins from the *Colored American Magazine*, January 1901. *Source:* James Weldon Johnson Memorial Collection in the Yale Collection of American Literature, Beinecke Rare Book and Manuscript Library, Yale University.

CLUB LIFE AMONG COLORED WOMEN. BY PAULINE E. HOPKINS.

THE
COLORED AMERICAN
MAGAZINE

· 15 CENTS A NUMBER   AUGUST, 1902 $1.50 A YEAR.

AN ILLVSTRATED MONTHLY DEVOTED TO LITERATVRE, SCIENCE, MVSIC, ART, RELIGION, FACTS, FICTION AND TRADITIONS OF THE NEGRO RACE.

MRS. BIRDIE HIGH,
St. Paul, Minn.
See page 299.

~ PVBLISHED BY ~
THE COLORED CO-OPERATIVE
PVBLISHING COMPANY
5 PARK SQVARE BOSTON MASS.

Cover of the *Colored American Magazine*, August 1902. *Source:* James Weldon Johnson Memorial Collection in the Yale Collection of American Literature, Beinecke Rare Book and Manuscript Library, Yale University.

In the magazine and her other publications, Hopkins sought to provide Black middle-class readers with historical connections to the continent of Africa, links that were often missing from the lives of the descendants of the enslaved. In 1905, she self-published from her Cambridge home a pamphlet called "A Primer of Facts: Pertaining to the Early Greatness of the African Race and the Possibility of Restoration by Its Descendants." In the pamphlet, she asked rhetorically (punctuation and capitalization hers):

> What is the obligation of the descendent of Africans in America?—To help forward the time of restoration [of advanced civilizations on the African continent]. HOW MAY THIS BE DONE?—By becoming thoroughly familiar with the meagre details of Ethiopian history, by fostering race pride and an international friendship with the Blacks of Africa.[81]

Hopkins often portrayed interracial relationships in her fiction, and she did so unapologetically. In response to one reader's letter, she wrote, "My stories are definitely planned to show the obstacles persistently placed in our paths by a dominant race to subjugate us spiritually. Marriage is made illegal between the races and yet the mulattoes increase. . . . Amalgamation is an institution designed by God for some wise purpose, and mixed bloods have always exercised a great influence on the progress of human affairs."[82] Knight says Hopkins's aim was to build public understanding for those of mixed race. "She understood what those of us who teach literature understand today, that literature helps people learn empathy," Knight says. When one reader said her characters should be less controversial and more like Charles Dickens's, Hopkins replied that "Dickens writes about real people and I'm doing the same thing," Knight adds.

Hopkins clashed with influential civil rights leader Booker T. Washington, whose attitudes about the role of African Americans in the country's economic life were at odds with hers. While Washington advocated for expanded economic opportunities for Black Americans, he stopped short of demanding full political rights. Hopkins was unwilling to accept the Jim Crow policies that effectively assigned African Americans to second-class status. She shared her vision of economic self-advancement through profiles of successful people in the magazine, titled "Famous Men" and "Famous Women." In asserting that success was within reach for Americans of color and offering role models, she was decades ahead of her time. Washington was among those she profiled, along with fellow Cambridge resident Maria Baldwin.[83] She also wrote about William Wells Brown, regarded as the first Black playwright in America and a Cambridge resident in the 1860s and 1870s. Brown was something of a mentor to Hopkins, sponsoring an essay contest that she won as a high school student in Boston.

Hopkins blamed Washington for a takeover of the financially struggling *Colored American Magazine* in 1904 that effectively ended her literary career. The new publisher moved the editorial offices from Boston to New York City. At first, she moved with it. As she recalled in a 1905 letter to William Trotter, a Boston newspaper publisher and civil rights champion, "I was offered $12 per week which I decided to accept having determined that I would accept the situation as I found it, succumb to the powers that were, and do all I could to keep the magazine alive."[84] Although "many promises were made me," she soon found herself "frozen out" by a nephew of Booker T. Washington's wife and was forced to resign in September 1904.[85] Although she tried several times to regain her platform, she was never again to have as large an audience and spent years struggling to reach any audience at all.

Pauline Elizabeth Hopkins was born in Portland, Maine, in 1859. Her family moved to Boston when she was a young girl, and there she attended Girls High School. After leaving school, she began a career as a stage performer. Hopkins wrote an original musical play entitled *Slaves' Escape; or, The Underground Railroad* that was performed when she was just twenty years old. It is believed to be the first play written by an African American woman and the first by an African American to be performed on the stage, Gruesser says, describing it as a bridge between minstrelsy and the Black theater that emerged around the turn of the twentieth century. In the 1880s, she performed with members of her family in a vocal ensemble, the Hopkins Colored Troubadours. Gruesser says she was often identified as "Boston's favorite colored soprano."

In 1895, Hopkins passed the civil service exam and became a stenographer for the Commonwealth of Massachusetts Bureau of Labor Statistics.[86] With the security of a government job, in 1896 Hopkins was able to purchase a newly built home in Cambridge, moving there with her mother and stepfather.[87] She left her state job in 1899 to begin a full-time writing career, a bold leap at the time.[88]

Hopkins was frequently an invited speaker at civic and women's club meetings in the Boston area.[89] She was active in civic organizations and women's clubs. In 1901, she was a founder, along with publisher William Monroe Trotter, of the Boston Literary and Historical Association, an adult education society.[90]

After leaving the *Colored American Magazine* in 1904, she wrote briefly for an Atlanta-based publication, *Voice of the Negro*, and continued her public speaking.

In 1916, she started the *New Era Magazine*, which briefly served as a forum for her ideas about racial progress, explored the history of Black America and its connections to the African continent, and celebrated accomplished men and women of color.[91] She had worked out a detailed plan for the *New Era*, lining up writers from Africa and

Historic plaque in front of Pauline Elizabeth Hopkins's former home on Clifton Street in Cambridge, 2020. *Source:* Photo by the authors.

elsewhere to contribute to future issues. She may have sold her house to fund the new magazine. Even so, she was unable to keep it going financially, and only two issues were ever published. "Of the many sad things [in her life], that's one of the most sad," Gruesser says.

The failure of *New Era* marked the end of Hopkins's more than forty years in the public eye. A combination of sexism plus a withholding of support from Booker T. Washington and his allies made a comeback impossible for her, Knight says. "Gender really does come into play when you think about the years afterwards—being a

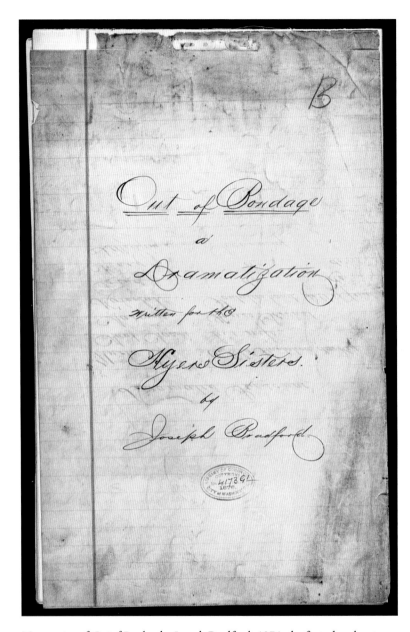

Manuscript of *Out of Bondage* by Joseph Bradford, 1876, the first play about slavery with an African American cast. Produced with Pauline Elizabeth Hopkins, Anna Madah Hyers, and Emma Louise Hyers. *Source:* Manuscript Plays Collection, Library of Congress Copyright Office.

woman in the field of journalism and literature more broadly," she says. W. E. B. Du Bois, whose work Hopkins published though there is no record of the two ever meeting, was able to survive his ideological battles with Washington, but she could not.

Research in recent years has revealed that Hopkins was, by modern standards, a plagiarist, Gruesser says. In the *Colored American*, writing serial novels, Hopkins often lifted passages from William Shakespeare, Alfred Tennyson, and the popular fiction of the day without attribution. Hopkins never acknowledged doing so nor explained in print or in any extant letters the motivations for her literary borrowings. She took perhaps as much as a quarter of the language of her novel *Hagar's Daughter* from other sources. "[It sometimes seems that] anything that serves her purpose she takes," Gruesser says. But her plagiarism, he argues, doesn't take away from Hopkins's importance: "it just makes her that much more fascinating."

Hopkins never married or had children. After being forced to sell her Clifton Street home in 1916, she moved into rented rooms on Jay Street in the Cambridgeport neighborhood.[92] She continued to work as a stenographer and from 1918 may have worked for a researcher at MIT, Gruesser says. She died at the age of seventy-one, after burning herself badly on an oil stove in her home. Today, an organization of literary scholars, the Pauline Elizabeth Hopkins Society, continues to explore her written work and legacy.

ROBERT LOWELL AND MODERN AMERICAN POETRY

A surprising number of the twentieth century's most influential poets lived, studied, or taught in Cambridge, including T. S. Eliot, e. e. cummings, Robert Frost, Robert Lowell, Stanley Kunitz, May Sarton, Elizabeth Bishop, Adrienne Rich, Seamus Heaney, and Louise Glück.

For much of the 1960s, a small seminar room beneath the dining hall of Harvard's Quincy House was a center of poetry writing in America. In two- or three-hour open workshops held weekly during the school year, Pulitzer Prize–winning poet Robert Lowell offered his opinions on poems presented by anyone willing to attend.[93] One week, Robert Pinsky, Frank Bidart, or Gail Mazur—each of whom later achieved national prominence as a poet—might hand out copies of recent work.[94] Other weeks, local residents who fancied themselves poets might participate, and occasionally, a homeless person seeking a warm place to sit would listen quietly. Bidart, who joined Lowell's writing seminar in 1966, recalled: "People would bring their poems and sit around a desk and pass them around and everybody would talk about them. You didn't have to be connected to Harvard at all. He welcomed anybody."[95] In the windowless, airless seminar room, Lowell would drag deeply on his ever-present cigarettes and offer critiques. "He didn't like most of the poems—either didn't get them or didn't like them or wasn't all that interested," Lloyd Schwartz, a professor of English at the University of Massachusetts in Boston, says. "But somehow there was always something profound that he had to say about poetry and about writing poetry and how this was the center of his life."

Robert Lowell at the Grolier Bookshop in Harvard Square in the 1960s. *Source:* Elsa Dorfman.

Lloyd Schwartz reading at the Grolier Poetry Festival on Plympton Street in Cambridge, 2018. *Source:* Photo by the authors.

Though hardly a household name, Lowell was considered "the greatest American poet of the mid-century, probably the greatest poet writing in English," according to one biographer, and "the most celebrated poet in English of his generation."[96] First at Boston University and later at Harvard, Lowell mentored well-known poets such as Sylvia Plath, Anne Sexton, and Elizabeth Bishop and influenced an entire generation of writers.[97] But at his Quincy House seminars, Lowell never talked about his own work, says Schwartz. "Everybody would have ached to hear him talk about his own poems, but he didn't."[98]

Lowell represented two major strains of twentieth-century poetry. His academic poetry was informed by writers such as Plato, Dante, Shakespeare, and Baudelaire. He was a voracious reader and interpreter of the poetry of the past and a scholar of Classical and nineteenth-century European poetry. Following the publication of *Life Studies* in 1959, Lowell was also recognized for his groundbreaking confessional poetry, using his troubled personal relationships as fodder for his writing. Some critics considered *Life Studies* the most important book of modern poetry since T. S. Eliot's 1922 *The Waste Land*.[99] Its informally composed autobiographical poems chart Lowell's search for some form of inner peace.

Throughout his life, Lowell suffered from what today is called bipolar disorder and spent many months in psychiatric hospitals in the Boston area, New York, and England.[100] In both his manic and depressive phases, he could be erratic, impulsive, abusive, and self-destructive, but he was nonetheless cherished by a circle of friends, lovers, former lovers, mentors, and students. Lowell came from one of Boston's most prominent families. He was related to nineteenth-century Cambridge poet James Russell Lowell, early twentieth-century poet Amy Lowell, and Harvard president A. Lawrence Lowell, but he seemed as much embarrassed and troubled by his family connections as uplifted by them. He never established a household in Cambridge but was a key part of its poetry community from the 1940s to the 1970s.

At Quincy House, an undergraduate dorm complex at Harvard, Lowell lived in a small apartment on the two or three days a week he spent in Cambridge, commuting from New York City.[101] For his office hours, Lowell had only a stuffy basement seminar room—"but he had a room and he took advantage of it," Schwartz says. After the workshops, Lowell and his acolytes often wandered over to a nearby Spanish restaurant to share a pitcher of sangria and continue their conversation. "It was part of the poetry life in Harvard Square."

A constellation of libraries, bookstores, coffee shops, literary societies, and arts venues in the Harvard Square area supported the poetry ecosystem. Poet Kathleen Spivack, who studied under Lowell, noted that in the literary world of the late 1950s and 1960s, "intellectuals sat in the all-night cafeterias discussing great thoughts. Most of the shops sold books, books, books: used ones and new ones and foreign ones. Small whiskered men darted in and out of these shops. . . . There were stationery shops, a typewriter store, and very little else that was not devoted to the life of the mind."[102] She compared Cambridge favorably to Paris. "The center of the universe! And Harvard Square our Montparnasse," she wrote. "We were living in an era of greatest American poetry, and these poets were living in our midst. The Cafe Pamplona was our Deux Magots; the Hayes-Bickford Cafeteria our Brasserie Lipp. Out of the steamy windows of the cafeterias we saw famous people—wasn't that Robert Frost?—hurrying, coat collars up, huddled against the cold."

Cafe Pamplona, 2018. Josefina Yanguas, who arrived in Cambridge from northern Spain after World War II, opened the basement-level Cafe Pamplona in this building on Bow Street in 1959. It is considered the first cafe in Harvard Square, a place where patrons could linger over a cup of coffee and good conversation. Cafe Pamplona was long a meeting place for poets, artists, and intellectuals. *Source:* Photos by the authors.

The Grolier Poetry Book Shop, a tiny Harvard Square storefront that opened in 1927, became a nexus for poets in the region and an important stop for visitors. The oldest operating poetry store in America, the Grolier became a place not only to purchase books of poetry but to socialize, read, and hear from favorite authors.

Harvard's Woodberry Poetry Room opened in 1931 to collect English-language poetry and audio recordings of poetry readings. The dedicated space signified Harvard's early acknowledgment of the importance of modern poetry.[103] T. S. Eliot's first poetry recordings were made at Harvard in the 1930s, pressed into vinyl LPs in 1933 for a record label called Harvard Vocarium. The Poetry Room, housed since 1949 in a space designed by Finnish architect Alvar Aalto, continues to foster recordings of poetry readings and convene symposia.

The *Harvard Advocate*, an undergraduate literary magazine established in 1866, has been an outlet for many important twentieth-century poets, publishing the undergraduate poetry of T. S. Eliot, Wallace Stevens, and e. e. cummings.[104] Other literary magazines flourished in Cambridge for a time in the 1960s and 1970s, Schwartz says, though most eventually folded. *Ploughshares*, begun in 1971 and named after the Plough and Stars bar on Massachusetts Avenue, may be the only survivor. It is now housed at Emerson College in Boston.

The Poets' Theatre, established in Harvard Square in 1950, offered live performances of poetry and poetic drama. Between 1951 and 1960, the Poets' Theatre staged more than seventy performances or readings, including works by Archibald MacLeish, William Butler Yeats, and Samuel Beckett. Writers who appeared live at the theater included poets Dylan Thomas, May Sarton, and Anne Sexton and novelist/playwright Truman Capote.[105] In 1960, the theater's home on Palmer Street in Cambridge burned to the ground, and its productions dwindled through 1968, when operations ceased altogether. The theater was briefly revived in 1986 and again in 2014.

Perhaps Cambridge's poetic life is less robust today than it once was, but there are still bookstores and cafes in Harvard Square, and poets continue to roam the streets. "There's [long been] some jealousy of the literary life in Cambridge," says Schwartz, who was named poet laureate in nearby Somerville in 2019, "and I think that's still true."

Prominent poets who have lived, studied, or taught in Cambridge since 1900 include the following:

- **T. S. Eliot** (1888–1965) wrote *The Wasteland* (1922), one of the most critically acclaimed English-language poems of the twentieth century. In 1909, Eliot received a bachelor's degree from Harvard after three years of study. Following a year in Paris, he returned to Harvard to pursue a PhD in Indian philosophy and Sanskrit, living in rented quarters on Ash Street.[106] He earned early notice with his modernist poem "The Love Song of J. Alfred Prufrock." Leaving Harvard in 1914 before finishing his doctoral degree, Eliot lived the remainder of his life in England, writing to his friend, fellow student and poet Conrad Aiken, "I dread returning to Cambridge . . . and the people in Cambridge whom one fights against and who absorb one all the same."

- **e. e. cummings** (1894–1962) was born and educated in Cambridge. His father taught sociology at Harvard before becoming a Unitarian minister.[107] The Cummings family lived on Irving Street near the home of psychologist William James. Cummings wrote poetry daily from the age of eight to twenty-two and penned nearly three thousand poems over the course of his life. He attended public schools in Cambridge—the Agassiz School, where Maria Baldwin served as principal, and then Cambridge Latin School—before graduating from Harvard in 1915. Cummings is noted for his use of free verse (poems lacking rhyme or meter), as well as the lack of capitalization and irregular punctuation in his poems. His satiric poem "the Cambridge ladies who live in furnished souls" includes the line "they believe in Christ and Longfellow, both dead." It is often studied in college literature classes. Cummings moved to Paris during World War I, serving as an ambulance driver. He later lived in New York, where he published his first book of poetry in 1923. He wrote twelve books of poems, one published posthumously.

- **Robert Frost** (1874–1963) is arguably the most popular American poet of the twentieth century. Frost made rural New England and the ordinary people who inhabit it the subject of many of his poems. He received four Pulitzer Prizes for poetry in his lifetime and recited a poem at the inauguration of President John F. Kennedy in 1961. Frost attended Harvard from 1897 to 1899 as an undergraduate, though he left due to illness before graduating. He lived his last twenty years, from 1943 to 1963, on Brewster Street in Cambridge, where he was visited by fellow poets and admirers, including Robert Lowell and Adrienne Rich. For aspiring writers in the Cambridge area, poet Peter Davison has written, Frost had no rival "as an example of what it meant to be a poet, to live as a poet, to think as a poet."[108]

- **Stanley Kunitz** (1905–2006) received his bachelor's and master's degrees in English at Harvard. First published in 1930, he received the Pulitzer Prize in poetry in 1959 for his *Selected Poems*. He went on to receive a National Book Award in 1995 and was named the US poet laureate in 1974 and 2000. Arriving in the Boston area in the fall of 1958 for a visiting professorship at Brandeis University, Kunitz took an apartment in Harvard Square, where he befriended poets Sylvia Plath and Ted Hughes.[109] Kunitz soon became a trusted poetic adviser to Robert Lowell; Lowell apparently

Cambridge homes of a few notable twentieth-century poets. Clockwise from top left: Robert Frost (Brewster Street), Elizabeth Bishop (Brattle Street), and Adrienne Rich (Brewster Street). *Sources:* Photos by the authors.

Former Cambridge homes of poets T. S. Eliot on Ash Street (top) and May Sarton on Channing Place (bottom). *Source:* Photos by the authors.

introduced him to Robert Frost one wintry afternoon in 1958 after lunch in Harvard Square.

- **May Sarton** (1912–1995), a prolific poet and novelist noted for her feminist poetry, lived in Cambridge as a child and teen. Her family fled Belgium in World War I, her father coming to teach at Harvard in 1916. Sarton enrolled at the alternative Shady Hill School in Cambridge, where her interest in poetry was encouraged, and graduated from the Cambridge High and Latin School in 1929.[110] A year later, she published her first poems, a series of sonnets, in *Poetry* magazine. Traveling extensively in Europe in the 1930s, Sarton published two volumes of poetry during the decade and began writing novels. Sarton's work has been read and studied widely, particularly in gender studies curricula, though she rejected the label of "lesbian writer" as limiting to her creative ambition. She sold her family home in Cambridge in 1958, moving to New Hampshire.

- **Elizabeth Bishop** (1911–1979) published her first book of poems in 1946 and won the Pulitzer Prize for poetry a decade later. A meticulous crafter of verse but never a prolific writer, Bishop published only a hundred or so poems in her lifetime, many relating to themes of belonging, abandonment, and grief.[111] She maintained a decades-long correspondence with Lowell and, at his insistence, was invited by Harvard to teach a course in poetry in the fall of 1970 (when he was too sick to teach it himself). Bishop held a teaching post in Harvard's English department for seven years, a rare role for a woman at the time. Bishop lived for a few years in the Brattle Arms apartments in Harvard Square and befriended poet Adrienne Rich, who was teaching at Brandeis University in nearby Waltham. Bishop wrote at least one poem about Cambridge. "Five Flights Up," the last poem of her final book, *Geography III*, describes a dog being scolded by its master, a scene she saw from her partner's fifth-floor apartment on Chauncy Street.[112] Bishop's reputation as a poet grew after her death at the age of sixty-eight.

- **Adrienne Rich** (1929–2012) was a widely read poet of the late twentieth century who became known for her politically infused poems decrying the treatment of women and lesbians in American society. Rich was not always a rebel: "Arriving at Radcliffe, the daughter of a Southern Protestant pianist mother and a Jewish doctor father, Rich initially excelled at being exceptional in accepted ways," according to poet Claudia Rankine.[113] She received acclaim in 1951 as a twenty-one-year-old student, following the publication of her first collection of poems, *A Change of World*. Rich married Harvard economics professor Alfred Conrad in 1953, settling on Brewster Street in Cambridge (not far from Robert Frost) and raising three sons. The pair regularly socialized with Robert Lowell, Sylvia Plath, and other Boston-area poets. Rich became increasingly dissatisfied with her roles as wife and mother and felt they had limited her opportunities in life. She turned this frustration with societal convention into her poetic work, most notably *Snapshots of a Daughter-in Law: Poems 1954–1962* published in 1963. The family moved to New York in 1966, where Rich became increasingly aligned with the antiwar movement, feminism, and other political causes.

- **Seamus Heaney** (1939–2013), a towering figure of modern Irish poetry and winner of the 1995 Nobel Prize in literature, lived and taught part-time at Harvard from 1979 to 2006.[114] Heaney wrote more than twenty volumes of poetry and criticism. His work, largely focused on the strife-torn territory of Northern Ireland where he was born, received both critical and popular acclaim. Arriving at Harvard after both Lowell and Bishop, Heaney became a center of Harvard's poetry community, Lloyd Schwartz says. "Seamus was pretty accessible and gave readings and was very generous." His arrival, Schwartz says, "was a real gift to the community" of poets and writers in the Boston area. As a visiting scholar, Heaney was offered a two-bedroom suite at the top of a winding flight of stairs in Harvard's Adams House on Plympton Street. The suite has since been dedicated in his name as a setting for quiet, creative work by Adams House's undergraduate residents.[115]

- **Frank Bidart** (1939–) moved from California to Cambridge in the early 1960s to study at Harvard and became a friend and student of both Robert Lowell and Elizabeth Bishop. He has taught at Wellesley College since 1972 and received the Pulitzer Prize for poetry in 2018 for *Half-Light: Collected Poems 1965–2016*.[116] Some of his best-known poems focus on the lives of troubled characters, including a child murderer and an anorexic woman. He has lived in the same Cambridge apartment for more than three decades, surrounded by stacks and boxes of books.[117]

- **Louise Glück** (1943–), appointed poet laureate by the Library of Congress in 2003 and awarded the Nobel Prize in literature in 2020, has lived in Cambridge for decades. Glück was only the third American poet to win the Nobel Prize—the first two were T. S. Eliot and Bob Dylan—and the third American female to claim the literature prize. Glück has written thirteen books of poetry.[118] The Poetry Foundation notes that she is known "for her poetry's technical precision, sensitivity, and insight into loneliness, family relationships, divorce, and death" as well as for her references to Greek and Roman mythology.[119] The chair of the Nobel Committee, in announcing Glück's award, cited her "biting wit" and a poem in which she references her connection to Cambridge: "I thought my life was over and my heart was broken. / Then I moved to Cambridge."[120]

SOCIAL REFORM IN THE PEOPLE'S REPUBLIC

The Puritan founders of the Massachusetts Bay Colony were so profoundly discontented with Stuart England that they were willing to risk their lives on a 3,200-mile sea voyage to an unfamiliar land. These early settlers were not simply economic migrants but were making a political and religious stand. They established their own self-administered churches, formed local governments without the interference of Mother England, and started their own educational institutions, all in an effort to create a new Eden, free from the corruptions of the Old World. It should come as no surprise, then, that Cambridge has been the home of political dissenters, challengers of the status quo, social reformers, and radical intellectuals for nearly four centuries.

One of the lessons we learned from researching this book, as detailed in the final chapter, is the degree to which innovators in Cambridge have been driven by a social mission or sense of moral obligation, not simply economic gain. From Cambridge's earliest days, the role of developing moral leaders was seen as central to its mission. Harvard College was formed in large part to educate the colony's next generation of ministers and political leaders, a role it continues to play on a national level.

A sense of moral purpose doesn't emerge only from the city's educational institutions. The people featured in this chapter all carried and acted on their own sense of duty and need for social reform. They envisioned and tried to bring about a better world, not just for themselves but for others. Cambridge native **Margaret Fuller** was a pioneering journalist and feminist, the first female foreign correspondent for an American newspaper. In 1845, decades before US women were granted the right to vote, she wrote "Woman in the Nineteenth Century," making the case for women's rights.

In 1831, the creation of **Mount Auburn Cemetery** on the Cambridge-Watertown border helped to reshape American attitudes toward death, grief, and the natural world.

Melusina Fay Peirce searched for ways to release American women from the drudgery of their domestic duties and unleash their greater potential. In 1869, she formed a cooperative in Cambridge to both streamline and assign economic value to housework.

Maria Baldwin, a beloved African American educator, was named the principal of a mostly white Cambridge public school in 1889. She hosted weekly literature and consciousness-raising meetings at her Cambridge home that included future founders of the NAACP.

W. E. B. Du Bois—the pioneering sociologist, civil rights advocate, and NAACP founder—moved to Cambridge in 1888 to further his education at Harvard. He honed his ideas about race and American society during this time and in 1895 became the first Black person to receive a PhD from Harvard.

The last vignette in this chapter explores the issue of **marriage equality**. The first legal same-sex marriages in the United States took place at Cambridge's City Hall in 2004.

The city has earned the nickname "the People's Republic of Cambridge" due to its reputation for left-leaning politics and support for social reform movements. Though intended as a derisive comment by political conservatives in the 1980s and 1990s, the nickname has been widely embraced by area residents. There was even a popular bar near Central Square called The People's Republik (1997–2021).

As chapter 9 addresses, idealism and passion for improving the world have characterized Cambridge for nearly four centuries and helped it thrive.

MARGARET FULLER: EARLY FEMINIST AND JOURNALIST

In July 1850, the Transcendentalist philosopher Ralph Waldo Emerson sent his friend Henry David Thoreau on a mission: to find whatever remains he could of the writer and women's rights advocate Margaret Fuller.[1] A few days earlier, Fuller had been shipwrecked off the coast of Fire Island, New York, while returning to America from Italy. Only forty at the time, Fuller was carrying the manuscript of a book she had written about her time as a war correspondent. Although nearly everyone else on the ill-fated ship survived, Fuller, her husband, and their two-year-old son did not—a tragedy that today seems almost inconceivable. Thoreau found little to report back to Emerson except a short list of their friend's belongings, including a metal chest with the letters MF painted over, perhaps in honor of her newly married status.[2]

Born in Cambridge in 1810, Sarah Margaret Fuller Ossoli became one of the leading writers of her time. She was a collaborator and close friend of Emerson's, serving in the 1840s as editor of *The Dial*, his short-lived but influential literary journal. Regarded as a pioneering feminist, Fuller in 1845 published *Woman in the Nineteenth Century*, a book—almost a manifesto—that influenced Susan B. Anthony, Elizabeth Cady Stanton, and the women's rights movement. At a time when the intellectual abilities of women were considered inferior, Fuller learned to read Latin, French, Italian, and German and translated a book by the German writer Johann Wolfgang von Goethe into

Margaret Fuller, head-and-shoulders portrait, facing left, created between 1840 and 1880. *Source:* Courtesy of the Prints and Photographs Division, Library of Congress, Washington, DC.

English. While few of her female contemporaries held paying jobs, Fuller served as a professional literary critic for Horace Greeley's newspaper the *New York Tribune*. In the last years of her life, she earned a living as a correspondent for the *Tribune*, covering the struggle for a unified and democratic Italy. She was the first American woman to work as a foreign correspondent.[3]

Fuller was born on Cherry Street in Cambridge near today's Central Square. Her father, Timothy Fuller, was a Harvard-educated, politically ambitious lawyer descended from early Puritan settlers of Massachusetts. Her mother, Margaret Crane, was a gentle and often passive presence who balanced Timothy's contentious nature. Timothy Fuller had high expectations for himself and his eldest child. He taught Margaret to read at the age of three, and after the death of a younger daughter, he focused even more of his attention on her education. When Margaret was seven, her father won a seat in the United States Congress and left the family for half of each year to serve in Washington. Later in life, Margaret was to recall her father as a tyrant who insisted on late-night recitations of her studies and as a harsh and unyielding critic whose approval and love she struggled to gain. One biographer noted: "Margaret came out of this training with a wide knowledge of history, of the works of Thomas Jefferson, of several languages, of English literature, of mathematics, and of Biblical scholarship, far beyond that possessed by the young men of her circle."[4] By her teenage years, Fuller was regarded by her family as intellectually precocious but socially inept. At age fifteen, she was sent by her father, firmly against her will, to a school for young ladies in Groton, Massachusetts. His hope was that she would develop the social graces that her early education at his hand had neglected. After a year of boarding school, Fuller returned home to continue her study of literary classics and modern languages. She read widely and voraciously.

Margaret Fuller House, Cambridge, ca. 1910–1920. *Source:* Courtesy of the Prints and Photographs Division, Library of Congress, Washington, DC.

In 1833, Timothy Fuller left his Boston law practice and moved his family, including twenty-three-year-old Margaret, from Cambridge to the town of Groton, drawn by its rural charm and a chance to start afresh as a gentleman farmer. For Margaret, being forced to leave her literate friends in Cambridge was a blow. Her troubles only grew on the death of her father from cholera in 1835, leaving the family with few assets or means of support. To help her ailing mother and provide for her siblings, Fuller took teaching positions at two progressive private schools, first in Boston and later in Providence, Rhode Island. In 1835, in Cambridge, she met Emerson, who was on the cusp of his fame as an essayist and champion of Transcendentalism, a distinctively American and highly individualistic philosophy.

Fuller was accepted among men like Emerson, Thoreau, and Nathaniel Hawthorne but also stood out because of her strong drive to succeed, according to Fuller biographer Megan Marshall.[5] "To state your ambition was not a good thing, but she would do that," Marshall says, adding that it was "particularly odd because there wasn't any clear way to achieve her ambition." Despite her strong personality, Fuller was not entirely unfeminine. Marshall says, "I think Fuller probably did dress very well and had good general self-presentation, except she'd be speaking to you the way a man would be speaking to you, and that was unsettling [to many men]."

The Fullers sold the Groton farm in 1839, and Margaret and her mother moved to Jamaica Plain, then outside Boston's city limits. When their lease expired in 1841, Fuller lived an itinerant life for a time, staying with friends in Cambridge, Concord, and Newport, Rhode Island for short stints. She set her mother up with a household in Cambridge, to which she would return over the next four years. In 1843, from her Cambridge home, she wrote a book, *Summer on the Lakes*, an idiosyncratic travel narrative detailing a journey she took to the Midwestern United States.

In 1839, Fuller agreed to serve as the first editor of *The Dial*, Emerson's literary journal. Neither her teaching nor her editing proved lucrative; she was, in fact, never paid for her work on *The Dial*. But through Emerson, Fuller entered the orbit of some of New England's most influential writers and philosophers, including Hawthorne, Thoreau, and Louisa May Alcott.

To supplement her income, Fuller began to host a series of "conversations," inviting intellectual Boston women to free-form discussions of the issues of the day. These meetings formed the nucleus of an 1843 article she wrote for *The Dial*. She began the process of expanding the article to book length in Cambridge, completing the draft in Fishkill, New York. Published in 1845—a full seventy-five years before American women were granted the right to vote—*Woman in the Nineteenth Century* is considered a seminal work of the American feminist movement. In it, Fuller argues for expanded opportunities for all people, drawing inspiration from the Enlightenment philosophy of Englishman John Locke and the struggle for freedom and self-determination of the American and French revolutions. The Declaration of Independence famously reads: "We hold these truths to be self-evident; that all men are created equal. . . ." Fuller asserted that this truth needed to include *all* men and *all* women, and she rejected what she saw as the hypocrisy of the slave-holding Founding Fathers. She argued for the end of African slavery and protested the marginalization of Native American populations. She lamented the limited prospects for women like herself, without access to legal rights, educational opportunities, and prospects for earning a respectable living. Women could even be sea captains if they wished, she argued.[6] Drawing on her extensive reading of Western literature, Fuller bolstered her arguments with references to the many strong women in history, from ancient Greece to Shakespearean England. If women were so central to the legends of earlier civilizations, surely they deserved a role equal to that of men in contemporary society, she wrote.

Woman in the Nineteenth Century sold well, Marshall notes, though it was roundly criticized by male reviewers.[7] Some women readers also took Fuller to task for criticizing marriage while remaining unmarried, Marshall says. But Fuller felt that her single status enabled her to make comments about the institution of marriage that no married woman could.

In the autumn of 1844, Fuller left Cambridge and her native New England for good, invited by Horace Greeley to serve as literary critic for his *New York Tribune*. In accepting the position, Fuller became one of the first female journalists in America and its first professional book reviewer. The next year, she departed for Europe as the *Tribune*'s only foreign correspondent. After visiting England, she traveled to Italy. In Rome, she met Giovanni-Angelo Ossoli, whom she later described as "a person of no intellectual culture," who may never have read a book all the way through.[8] "Nature has been his book," she wrote. The two were apparently married (though the historical record is inconclusive on this point) after they conceived a child, whom they named Angelo after his father and called Nino.[9] Despite the social conventions of the times, Fuller did not try to pretend that she had married before becoming pregnant and suggested, according to Marshall, that she would have remained single but for the sake of the child.[10] "That was the thing that was most scandalous," Marshall says.

In 1850, Fuller and Ossoli decided to journey to the United States. With money tight, they chose to travel as inexpensively as possible, making the passage aboard a cargo vessel. Though her remains were never located, memorials were erected in her honor at Mount Auburn Cemetery and on Fire Island. Her Cambridge birthplace, now named the Margaret Fuller House, serves as a neighborhood social services center.

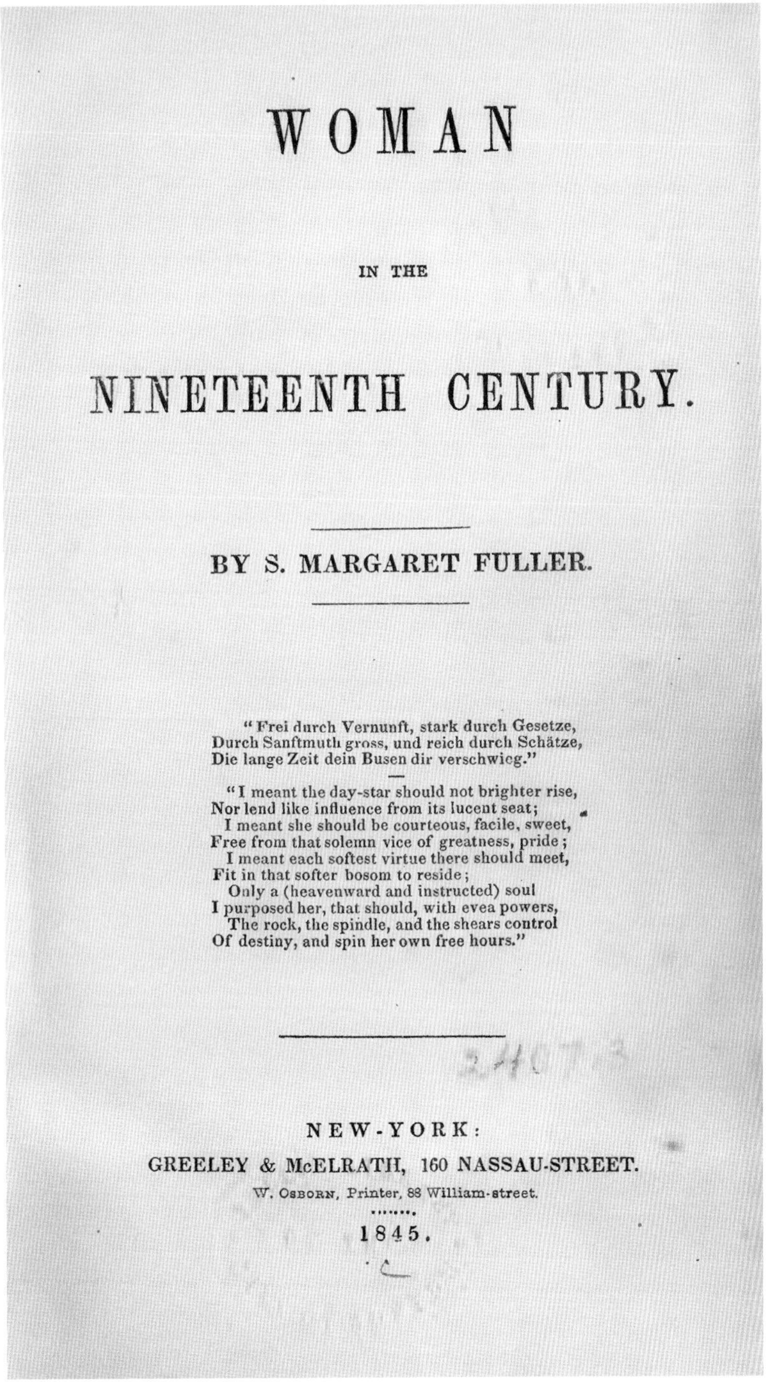

WOMAN

IN THE

NINETEENTH CENTURY.

BY S. MARGARET FULLER.

"Frei durch Vernunft, stark durch Gesetze,
Durch Sanftmuth gross, und reich durch Schätze,
Die lange Zeit dein Busen dir verschwieg."

—

"I meant the day-star should not brighter rise,
Nor lend like influence from its lucent seat;
I meant she should be courteous, facile, sweet,
Free from that solemn vice of greatness, pride;
I meant each softest virtue there should meet,
Fit in that softer bosom to reside;
Only a (heavenward and instructed) soul
I purposed her, that should, with evea powers,
The rock, the spindle, and the shears control
Of destiny, and spin her own free hours."

NEW-YORK:

GREELEY & McELRATH, 160 NASSAU-STREET.

W. Osborn, Printer, 88 William-street.
....
1845.

Title page of *Woman in the Nineteenth Century* by S. Margaret Fuller, 1845.
Source: Courtesy of Boston Public Library.

Cenotaph to Margaret Fuller Ossoli, Mount Auburn Cemetery, Cambridge, 2018.
Source: Photo by the authors.

Marshall says she thinks it's a tragedy that high school and college students rarely learn about Fuller along with classic New England authors and early feminists. Perhaps it's because her writing doesn't fit easily into the curriculum and she died before the suffrage movement took off. Fuller was unusual in her versatility as a writer; she could write travelogues as well as philosophical essays and reported pieces on what we'd call social justice issues. "Critics have been dismissive of her writing, which I think is wrong," Marshall says. "She was an innovator as a journalist."

MOUNT AUBURN CEMETERY

Cemeteries are for the living as much as for the dead. On a visit to Mount Auburn Cemetery on a late spring day, it's striking how alive the place is—with the sounds of birds, the budding green leaves of its oaks and beech trees, the murmurs of visitors, the smell of lilacs. The history of Boston can be read in the chiseled stone memorials to Lowells, Lawrences, and Longfellows that line the cemetery's wooded and gently rolling paths.

But many visitors fail to realize what a radical departure Mount Auburn Cemetery represented when it was established in 1831. Before Mount Auburn, America's dead were buried in crowded churchyards or on family homesteads. The nation's first so-called garden cemetery, Mount Auburn represented a changing attitude toward death. It helped birth the romantic notion of connecting people to their ancestors through the natural world and provided tranquility and restfulness that didn't exist in the urban burial grounds that preceded it. The cemetery, which straddles the town line between Cambridge and Watertown, became a popular destination in its early decades, acquiring an international reputation and rivaling Niagara Falls, the Erie Canal, and Mount Vernon as tourist destinations.[11] Charles Dickens visited the cemetery during a trip to the United States.[12] It was so popular with visitors that it was, for a time, closed to the general public on Sundays, so family members and friends of the deceased could visit their dead in relative peace.[13]

Mount Auburn's layout resembles that of a quaint suburban town, with winding paved roads that follow the contours of the land, grassy walking paths mostly named for botanical species, and stone monuments in neat groupings aligned to the paths. Most of the cemetery is covered by a majestic canopy of mature trees, which reveals itself in layers of color and texture and shadow. A natural depression in the landform—named the Consecration Dell—is the result of glacial activity millennia ago. Above the dell sits the highest point of the cemetery, where a sixty-foot-tall granite tower dedicated to George Washington was built in 1854; it predates the more famous Washington Monument in the nation's capital by thirty years.

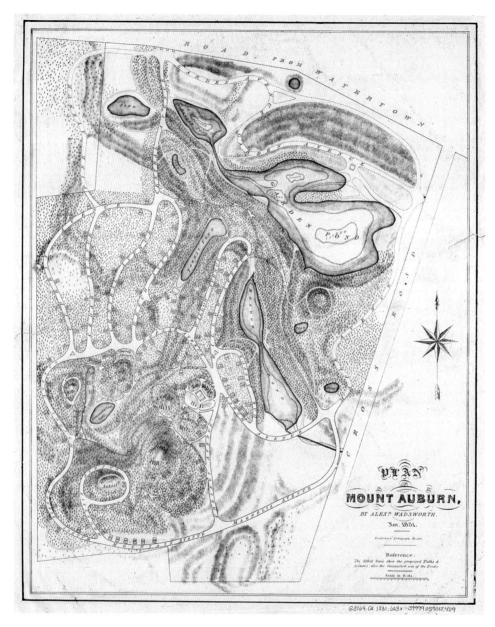

Plan of Mount Auburn Cemetery by Alex Wadsworth, November 1831. *Source:* Norman B. Leventhal Map & Education Center, Boston Public Library.

As a man-made landscape intended for the peaceful contemplation of nature, Mount Auburn preceded New York's Central Park, often heralded as the greatest of American urban parks, by twenty-five years. (It's not clear whether Frederick Law Olmsted, the designer of Central Park, ever visited Mount Auburn Cemetery, though he had certainly seen pictures.) Its creation inspired the design of other rural cemeteries, such as Philadelphia's Laurel Hill Cemetery (1836) and New York's Green-Wood (1838), and set off a wave of cemetery creation in nineteenth-century America.[14] Within its 174 acres, active recreation is not permitted, so unlike a public park, there is no bike riding, ball tossing, or picnicking. The experience of visiting is intensely serene. On a typical weekend day, the cemetery is populated with bird-watchers, nature lovers, history buffs, and the occasional mourner.

Many of Boston's and Cambridge's most prominent citizens have been buried here over the last two centuries, including a number of people profiled in this book: poet Henry Wadsworth Longfellow, bookseller and author John Bartlett, scientist Louis Agassiz, telescope maker Alvan Clark, botanist Asa Gray, photography pioneer

Engraving of Mount Auburn Cemetery, 1838. Artist: William Henry Bartlett, engraver: R. Brandard. *Source:* Collection of the authors.

Edwin Land, and physician Benjamin Waterhouse. Other Cambridge notables buried here include Clement Morgan, a lawyer and civil rights activist who helped found the NAACP, and Joyce Chen, the Chinese American chef who popularized Mandarin cuisine in America. The cemetery serves as the final resting place of others who achieved lasting fame: abolitionist and slave narrative author Harriet Jacobs, artist Winslow Homer, cookbook pioneer Fannie Farmer, physician and author Oliver Wendell Holmes Sr., architect Buckminster Fuller, Christian Science founder Mary Baker Eddy, and hospital reformer Dorothea Dix. The first burial at Mount Auburn, a stillborn infant, took place in July 1832.

The creation of the cemetery was driven in part by concern for public health. Dead bodies are vectors for disease, and as bodies piled up in the graveyards of Boston churches, they were seen as a health hazard. Inhumation—the burial of bodies in the earth—was preferred by early Americans, particularly members of Christian faiths, over cremation. Colonial-era burial grounds were purely functional spaces, unattractive and often poorly maintained. Burials were not guaranteed to be permanent in space-constrained urban graveyards. The spacious grounds of Mount Auburn Cemetery promised permanent commemoration of the dead.

In 1831, a half century after the nation's founding, the United States was also contending with the deaths of its early leaders. Men who were heroes at the time of the American Revolution were passing on, and the next generation was forced to consider its own legacy. Mount Auburn's creation was motivated in part by ancestry and hero worship and in part as a means to help forge a new vision for America, says Bree Harvey, vice president of cemetery and visitor services at Mount Auburn Cemetery. "You see it all coming together here," Harvey says. "They felt very strongly that they needed to help shape a new American identity." Emblematic of that democratic vision, the cemetery has many memorials dedicated by friends of the deceased, says Meg Winslow, curator of historical collections.

An equally significant reason for the cemetery's founding was the growing wealth of the Boston region in the early decades of the nineteenth century, as the city's shipping and manufacturing activities generated riches and created a wealthy upper class. Mount Auburn became a place to honor these new fortunes and their makers with grand permanent monuments.[15] Author Blanche Linden has noted that "rapid acceptance of Mount Auburn by the elite ensured the success of the venture."[16]

Credit for the initial conception and execution of Mount Auburn Cemetery belongs with two men: Henry Dearborn, a general during the War of 1812, founder of the Massachusetts Horticultural Society, and a US Representative from Massachusetts from 1831 to 1833; and Jacob Bigelow, a physician and professor of medical botany at

Harvard.[17] It was Bigelow who first suggested acquiring a plot of rural land for burial use in 1825, though it took several years to find and buy a suitable property. Dearborn thought that selling cemetery plots could fund his dream of an experimental garden to educate New Englanders about horticulture. He was responsible for the initial layout of avenues and paths—essentially, the landscape gardener for the cemetery—while Bigelow designed the Egyptian Revival gateway that marks the Mount Auburn Street main entrance, as well as a chapel and the Washington Tower.[18]

Bigelow had suffered the death of his firstborn son at the age of six months to "lung fever" in 1820. Joseph Story, the cemetery's first president, lost his ten-year-old daughter to scarlet fever. There were not many socially acceptable ways for a man to publicly express sorrow in early nineteenth-century America.[19] Both men sought to create in Mount Auburn a place where one could spend time in nature contemplating mortality.

Egyptian Revival entrance gate to Mount Auburn Cemetery, Cambridge, ca. 1890–1901. *Source:* Detroit Publishing Company photograph collection, Library of Congress.

Mount Auburn Cemetery, 2009. *Source:* Photo by the authors.

Story—a US Supreme Court Justice, Harvard Law professor, and an important early interpreter of the US Constitution—gave the cemetery's consecration address to a crowd of about two thousand:

> Thus, these repositories of the dead caution us, by their very silence, of our own frail and transitory being. They instruct us in the true value of life, and in its noble purposes, its duties, and its destination. They spread around us, in the reminiscences of the past, sources of pleasing, though melancholy reflection.
>
> We dwell with pious fondness on the characters and virtues of the departed; and, as time interposes its growing distances between us and them, we gather up, with more solicitude, the broken fragments of memory, and weave into our very hearts, the threads of their history. As we sit down by their graves, we seem to hear the tones of their affection, whispering in our ears. We listen to the voice of their wisdom, speaking in the depths of our souls. We shed our tears; but they are no longer the burning tears of agony. They relieve our drooping spirits. We return to the world, and we feel ourselves purer, and better, and wiser, from this communion with the dead.[20]

In 1860, Albert Edward, Prince of Wales and the future King Edward VII, came to Canada and the United States on a goodwill tour. Nearing the end of his months-long visit, the eighteen-year-old prince lunched at Harvard and then stopped at Mount Auburn, where he planted a European beech tree and an oak native to North America, Harvey says. "The fact that this is the spot that was chosen for that ceremony to take place says something about the significance of Mount Auburn, not just in Cambridge or Boston but on a national, an international scale," she says.

Today, Mount Auburn Cemetery remains an active burial place and tourist attraction, with more than 250,000 visitors a year.

MELUSINA FAY PEIRCE AND COOPERATIVE HOUSEKEEPING

In 1855, a young Melusina Fay, visiting her aunt in Cambridge, wrote a letter to her mother back home in Saint Albans, Vermont. Melusina suggested the family purchase a sewing machine. Her mother responded: "I sympathized with your feelings about the sewing machine very keenly; because every day I am easy till I begin to sew, & as soon as I sew the remorseless pain of my chest begins and increases so that I can't help believing the sewing to be one of the main causes of my suffering."[21] The family soon acquired a machine, possibly a gift from Fay's wealthy aunt, though not in time to help her mother.[22]

When Emily Hopkins Fay died the next year, she left Melusina, her eldest daughter, in charge of the household and five younger siblings. Emily had been cursed with poor health—possibly tuberculosis—and found herself overwhelmed by her duties as a minister's wife, mother, housekeeper, and schoolteacher.[23] Her death at age thirty-nine made a strong impression on twenty-year-old Melusina, who was convinced that the heavy burden of domestic duties had shortened her mother's life and kept her from realizing her potential. She resolved "never to cease until I saw some better way for women than this which can so horribly waste and abuse their finest powers."[24]

Melusina Fay Peirce (her full name after marriage) took up the reform of women's household labor as her lifelong quest and first put it into practice during her years living in Cambridge. Author Dolores Hayden has called Peirce one of "the first feminists in the United States to identify the economic exploitation of women's domestic labor by men as the most basic cause of women's inequality."[25] In an interview, Hayden explained that Peirce "wanted to change women's material conditions"—the physical structure of their homes and their domestic responsibilities—as a step toward equality with men.[26] Though gender roles have changed dramatically in America since the 1860s, the issues that Peirce raised continue to resound with women today.

Born in Vermont in 1836, Peirce (pronounced "Purse" by Bostonians) was the daughter of an Episcopal priest. Her mother raised seven children and oversaw girls' schools in the communities in which they lived—St. Albans, Vermont; Montpelier, Georgia; Bayou Goula, Louisiana; and New Orleans. The schools were a family affair. Peirce was expected to help teach the pupils and care for her younger siblings. As well as reading, writing, and arithmetic, she learned Latin, French, drawing, and music. Growing up in a church family, Peirce was held to high standards of moral behavior and was expected to contribute to larger society, imperatives with which she often struggled as an adult.[27]

Like the feminist author Margaret Fuller but twenty-six years her junior, Peirce was part of an upper-middle-class New England family that placed strong emphasis on the education of girls. Both inhabited a society that expected women to devote themselves to raising children and keeping households. The tensions inherent in these two directives—to be well-educated and literate but at the same time limit oneself to domestic life—played out across Peirce's own life. She would become a published author, school teacher, lecturer, and feminist organizer, as well as the house-tending wife of a Harvard-educated logician and philosopher.

On her father's side, Peirce was related to Judge Samuel Fay, whose family home overlooked the Cambridge Common, close to the spot where George Washington first took command of the Continental Army. In 1854, at the age of nineteen, Peirce went to live with her Aunt Maria Fay in Cambridge, thrilled to be in the company of so many bright minds and determined to study German. "Cambridge with its close intellectual

Portrait of Melusina Fay Peirce, 1876, at about age forty. *Source:* Dolores Hayden, *The Grand Domestic Revolution: A History of Feminist Designs for American Homes, Neighborhoods, and Cities* (Cambridge, MA: MIT Press, 1981), 66.

and liberal society of neighbors within walking distance of each other made possible and acceptable this ideal spinster existence," she said.[28] "Society was easy, and seldom stuffy. [Aunt] Maria is shelling peas when Mr. Longfellow and Mr. Lowell come to call. They sit down to help her and to talk—brilliant talk."

Peirce's plans to educate herself came to nothing until four years later, when she returned to Cambridge to attend the young ladies' school operated by Louis Agassiz and Elizabeth Cary Agassiz. Peirce spent two years studying in Cambridge, returning home in summers to St. Albans. It was in the Agassiz school that Peirce learned to apply rigorous scientific methods to the problems of the natural world and human society. Peirce wrote of herself: "the intellectual horizons that opened upon her from the lectures of Professor Agassiz and the lessons and lectures of the assistants, most of whom were Harvard professors like himself, were a revelation."[29]

By 1861, Melusina Fay was looking for ways to become financially independent, no small hurdle for a single middle-class American woman in the nineteenth century. She aimed to become a writer, hoping to publish essays on music in *Harper's* magazine. She began a courtship with Charles Peirce, the son of mathematics professor Benjamin Peirce and himself a student at Harvard. In October 1862, the two married in Vermont.[30] Returning to Cambridge, the newlyweds joined the Peirce family home on Quincy Street while Charles worked as a mathematician and sought a permanent academic position. Moving into rental quarters on Mount Auburn Street soon after their marriage, the young Peirces managed their expenses carefully on Charles's small salary from the US Coast Survey, with Melusina having no income of her own.

Peirce continued to support her husband's career and manage the household, which in 1864 included a home of their own on Arrow Street. "She ran the house for herself and Charley and whichever sisters were in residence, and at the same time she was considering, criticizing, trying to find ways to reshape the domestic scene. Everything she did from day to day—looking at the grease spots on the kitchen floor, smelling the 'swill pail'—must have fed her speculations. . . ." Sylvia Wright Mitarachi, Peirce's grandniece, wrote in an unpublished biography.[31]

Eight years after her mother's death, Peirce wrote a letter to the *New York World* newspaper on the predicament faced by American women. She noted that innovations like the power loom, which automated the production of cloth, had replaced household duties that traditionally fell to women. In her letter, Peirce suggested that middle-class American women, freed from mundane domestic tasks, could take on paying jobs rather than waste their days with society gossip and needless adornment.[32] Peirce described a society in which women could be wives and mothers while also conducting valuable business and earning incomes of their own, independent of a father or husband. Peirce believed that all types of human labor, even domestic work, merited compensation, making her perhaps the first American woman to publicly call for paid housework.

In 1868 and 1869, Peirce published a series of articles in the *Atlantic Monthly* magazine, advocating what she called "cooperative housekeeping." Peirce proposed that groups of twelve to fifty women organize cooperative associations, collectively meeting each member family's laundry, cooking, and other domestic needs.[33] "Laundry would be done in a central facility; cooked meals delivered to members' homes; and they would buy their supplies in a co-operative store where they would pay retail prices."[34] By pooling their resources, individual matrons would no longer require servants in their homes. While she was inspired by cooperative associations that then existed in England, her feminist take on the subject appears uniquely her own.

Peirce's ideas on domestic reform found an appreciative audience, and letters to the editor came in from across the United States. She was equally well known in England, where a British journal reprinted extracts of her articles and her ideas were published in a book.[35] She received requests for advice on starting cooperatives from writers as far away as Detroit, Michigan, and Crawfordsville, Indiana. She lectured on the topic of cooperative housekeeping to groups that expressed interest, such as the Fourth Women's Conference in Philadelphia in 1876 and the Illinois Social Science Association in 1879.[36] In 1884, Peirce published *Cooperative Housekeeping: How Not to Do It and How to Do It*.

Peirce's novel domestic ideas took tangible form with the establishment of the Cambridge Co-operative Housekeeping Society, led by Peirce and supported by a coterie of her Cambridge friends and family members. On June 10, 1869, nearly a hundred women filled the back room of the Cambridge Post Office to hear Peirce describe cooperative housekeeping, and on October 5, thirty-five women and two bachelors signed on to her project.[37] The society first met on May 6, 1870, with fourteen attendees.[38] Her speech before a group of Cambridge women in mid-June was printed in the local newspaper and distributed to potential members. She sought fifty subscribers to a proposed laundry at $50, fifty subscribers to a bakery at $25, a hundred subscribers to the laundry at $25, and twenty-five subscribers to the kitchen and bakery at $100.

With proceeds from the subscription drive, the society rented a building on Bow Street. Peirce oversaw the creation and operation of the laundry, though she was exhausted by this commitment. The laundry lost money in its first months under her leadership. Meanwhile, the kitchen was completed, and the store was stocked and opened to its members. In response to pleas from her husband, in October 1870 Peirce left for Europe to support him on a scientific expedition.[39] When she returned to Cambridge in the spring of 1871, she was filled with new energy and ideas to make the association a success, but by that point, her fellow members had largely abandoned the endeavor. The cooperative, which continued to operate at a loss in her absence, was dissolved by its officers in March 1871. Peirce must have felt extremely let down by the friends and neighbors with whom she had begun this grand experiment.

Peirce left her husband in 1876, and the two officially divorced in the early 1880s. For the remainder of her life, she devoted herself to one venture after another: supporting the education of women, helping to establish what became Radcliffe College, advocating for women to serve as deaconesses in the Episcopal church, writing classical music reviews for the *Chicago Evening Journal*, and leading a charge to restore New York

PROSPECTUS

OF THE

Cambridge Co-operative Housekeeping Society.

WE, the undersigned, women of Cambridge, have associated ourselves together, with the consent and hearty good-will of our families, for the purposes of purchasing their food and clothing at more reasonable prices, and of getting their cooking, laundry-work, and sewing, done better and more conveniently than at present.

To these ends, we propose to fit up this year a Kitchen and Laundry large enough to serve at least fifty households ; and next year to open Sewing Rooms which will furnish the clothing of these households.

To start the Kitchen and Laundry we need a capital of $10,000. We propose to raise the greater part of this sum by subscriptions of $100 for the Kitchen, and of $50 for the Laundry, payable cash down, or in instalments of not less than $10 per month. Subscriptions to the Store-room and Bakery belonging to the Kitchen will also be received, at $25 each. The Society will pay 7 per cent interest on the subscriptions, but no subscription will begin to draw interest until it is all paid in.

$3,000 are already pledged ; as soon as $5,000 are secured, steps will be taken to obtain premises suitable for the immediate purposes of the Society.

In 1866 the Legislature of this State passed an Act for the benefit of Co-operative Associations among workmen, by complying with the conditions of which, no member of such an association can be *personally liable* for any of its debts. It is the intention of this Society to apply for a General Act, equally favorable to Co-operative Housekeeping Associations, as soon as the Legislature meets in January, 1870 ; and until such Act has been obtained, and the organization of the Society under it completed, no subscriber will be required to pay his or her subscription, or any instalment of it, excepting a small assessment of two per cent for the preliminary expenses of the Society, such as printing, postage, &c. ; and no authority will be given to any agent of the Society to incur any liability not covered by cash in the Treasury appropriated for that purpose.

With these assurances, it is hoped that all who have hitherto been deterred from subscribing by fear of being made liable in case of failure, for more than the amount of their shares, will come forward at once, and join us, as we wish to get into working order, if possible, before the coming winter.

The business of the Society will be conducted on a strictly *cash* basis. It will not sell to its members at *cost*, but [according to the better principle established by the Rochdale Equitable Pioneers] at the current retail prices ; then, in each department, after paying expenses, and the interest on the shares, all the remaining profits of the business of that department are to be divided at stated times among the several subscribers to it, proportionately to the amounts of their purchases from it during the period in which such profits were accumulated.

Prospectus of the Cambridge Co-operative Housekeeping Society, 1869. *Source:* Courtesy of the Schlesinger Library, Radcliffe Institute for Advanced Study, Harvard University.

Home of the Cambridge Co-operative Housekeeping Society, 10 Arrow Street, Cambridge, 2020
Source: Photo by the authors.

City's historic Fraunces Tavern. To support herself and leverage her interest in domestic innovation, Peirce ran an upscale boarding house in Chicago in the early 1880s and another in New York City later in the decade. She also continued to write for newspapers in Boston, New York, and Chicago. Peirce lived to the age of eighty-seven, dying in a nursing home in Watertown, Massachusetts, in 1923.

Peirce was often overlooked in her lifetime and nearly forgotten later. At age eighty-two, she complained that social reformer Charlotte Perkins Gilman and others stole her ideas.[40] Peirce and her contributions remained largely unknown until Hayden came across her work in the late 1970s while doing research at Harvard's Widener Library. Today, "I think we have seen her influence in many, many countries of the world and many different kinds of experiments and programs," Hayden says, citing food and laundry service delivery as examples. "I think she launched those issues in a very powerful way."

Melusina Fay Peirce

Inventor

Melusina Fay Peirce's signature from her patent application for a dwelling block design, granted as US patent number 734938 in July 1903. *Source:* US Patent and Trademark Office.

MARIA BALDWIN: EDUCATOR AND REFORMER

The house on Prospect Street in Cambridge has looked forlorn for some time. As of mid-2021, its brown shingles remained worn, its small front garden untended. Yet this half of a Greek Revival duplex has been designated a National Historic Landmark in honor of its former owner: school principal and civil rights champion Maria Baldwin.[41] From 1889 to 1922, Baldwin served as the head of Cambridge's Agassiz School, where she was beloved by students and faculty and noted for her progressive approach to teaching. Baldwin hired the city's first school nurse, created the first open-air classroom and the first parent-teacher group in the district, and designed new ways of teaching math and science.[42] She was also a renowned public speaker, delivering addresses across the eastern seaboard on the lives of notable Americans, including George Washington, Abraham Lincoln, and Harriet Beecher Stowe.[43]

What made Baldwin truly remarkable was that she accomplished all this as a Black woman during an era widely regarded as a low point in race relations in America.[44] She was the first African American to lead a mostly white school in all of New England—with a dozen white teachers and nearly five hundred white students under her leadership.[45] "Maria Baldwin's distinguished career at the school provided a nationally known example of the abilities of black women educators," according to a 1976 National Park Service report establishing the house's landmark status. In 1917, sociologist and civil rights activist W. E. B. Du Bois honored Baldwin in the journal *The Crisis* in his "Man of the Month" column. He noted that "Miss Baldwin . . . without a doubt, occupies the most distinguished position achieved by a person of Negro descent in the teaching world of America, outside cities where there are segregated schools."[46]

Biographical information about Baldwin is limited, mostly consisting of short profiles that repeat the same information.[47] What we know is that she was born in Cambridge in 1856, the daughter of Peter Baldwin, a postal worker of Haitian descent, and

Portrait of educator and civic leader Maria "Molly" Baldwin, taken in the Boston photography studios of Elmer Chickering, ca. 1885. *Source:* Courtesy of the Prints and Photographs Division, Library of Congress, Washington, DC.

cent have continued their studies, after graduation from the school, in higher institutions of learning.

The purpose of Miss Miner to train her pupils in the art of living, as well as of teaching, is still a vital thing in Miner Normal School. The institution is contributing powerfully to the efficiency of family and school life, not only in the District of Columbia, but in many parts of the United States.

卐 卐 卐 卐 卐 卐

Men of the Month

A MASTER EDUCATOR

MISS MARIA F. BALDWIN was born in Cambridge, Mass., and trained in the public schools and the Normal School of that city. Her father was of West Indian descent and, during his early life, was a seaman; afterward he was for a long time employed in the city post office. Finding no employment at home Miss Baldwin first taught in Maryland but agitation of the colored leaders, during the time of Mayor Fox, led to her being appointed as teacher of primary grades in the public schools of Cambridge, in 1882.

In 1889, after teaching in all the grades, from the first to the seventh, Miss Baldwin was made Principal of the Agassiz School and retained that position for twenty-four years. In April, 1915, this school was torn down and a new building erected at a cost of $60,000, not including furniture.

MISS MARIA F. BALDWIN.

October, 1916, Miss Baldwin was made Master of the new Agassiz School, a position of great distinction, as there are but two women masters in the city of Cambridge. The school, composed of kindergarten and eight grades, is one of the best in the city and is attended by children of Harvard professors and many of the old Cambridge families. The teachers under Miss Baldwin, numbering twelve, and the 410 pupils, are all white.

Miss Baldwin thus, without doubt, occupies the most distinguished position achieved by a person of Negro descent in the teaching world of America, outside cities where there are segregated schools.

AN HONORED TEACHER.

DR. R. S. LOVINGGOOD, the president of Samuel Huston College in Austin, Tex., has passed on. He was born fifty-three years ago and his end came quietly

In April 1917, Maria Baldwin was profiled in the "Men of the Month" feature of the NAACP's publication, *The Crisis*, edited by her friend W. E. B. Du Bois. *Source:* The Modernist Journals Project, Brown and Tulsa Universities, modjourn.org.

Mary (Blake) Baldwin. She graduated from the city's high school in 1874 and from the Cambridge Teachers' Training School the next year. Unable to find suitable work in the Boston area, Baldwin began a teaching career in Maryland but returned to Cambridge in 1881 or 1882 (sources vary), accepting a job as a primary school teacher at the Agassiz Grammar School a few blocks north of the Harvard campus. First, she was given an "overflow" class and the following year was appointed to a full-time job, earning $750 a year.[48] After teaching each of the grades from one to seven, she was selected to serve as principal of the school in 1889.

In 1916, the Agassiz school was rebuilt and expanded to eight grades. Baldwin was influential in shaping the architectural plans for the building, insisting on open-air classrooms, an assembly hall, and a science museum.[49] Following completion of the building, Baldwin was named master of the school, a title of even greater distinction than principal.[50] The population of more than four hundred students was among the most privileged in the city, including many children of Harvard professors.[51] She was respected and beloved by her faculty, her students, and members of the local community. The poet e. e. cummings, a Cambridge native, wrote of her: "Never did any demidivine dictator more gracefully and easily rule a more unruly and less graceful populace. Her very presence emanated an honor and a glory."[52]

Baldwin also led another life, away from the largely white world of her school, according to her biographer, Kathleen Weiler.[53] She was a strong advocate for racial equality and social justice and an important member of Boston's African American intelligentsia at the turn of the twentieth century. She held weekly literary gatherings for Black students at Harvard, including Du Bois.[54] Her home at 196 Prospect Street became the headquarters for several clubs and social organizations, including the Omar Khayyam Circle, an elite Black literary society. Careful to avoid unwanted attention from those opposed to racial equality, many early civil rights advocates gathered under the guise of social, literary, or religious organizations.

Although no direct evidence remains, conversations held in Baldwin's house may have helped spark the Niagara Movement, the civil rights organization formed in 1905 by Du Bois, Boston newspaper editor William Monroe Trotter, and others. The Niagara Movement's members opposed the approach to racial progress promoted by activist Booker T. Washington. While Washington focused on economic self-improvement and better educational opportunities for Black Americans, Du Bois and his network of like-minded reformers insisted on fully equal civil and political rights. Trotter, an influential African American businessman and later cofounder of the National Association for the Advancement of Colored People (NAACP), was a frequent guest at Baldwin's home, as was Harvard-educated attorney and civil rights advocate Clement Garnett

The Maria Baldwin House on Prospect Street in Cambridge was named a National Historic Landmark in 1976. It was Baldwin's home when she served as the principal of the Agassiz School in Cambridge. *Source:* Courtesy of the Historic American Buildings Survey, Library of Congress, Washington, DC.

Morgan. Morgan was chosen by Du Bois as the Massachusetts state secretary of the Niagara Movement and later was active in the formation of the NAACP.[55] Another guest at Baldwin's home and participant in the Omar Khayyam Circle was William H. Lewis, a Harvard football legend and later assistant US attorney, the first African American to hold such high office in the federal government.

Baldwin herself was active in social and cultural organizations in the Boston area, especially those related to education and civil rights. She was appointed to the Committee of Forty, a group identified just after the first meeting of the NAACP to organize its founding—though Weiler says Baldwin was included largely as a figurehead. "There was a fight about putting on radicals," Weiler says, and Baldwin was seen as an alternative to more militant figures like journalist Ida B. Wells. As a sign of the respect

The Maria L. Baldwin School on Oxford Street in Cambridge, 2020. The former Agassiz School was renamed for Baldwin in 2002. *Source:* Photo by the authors.

accorded her in Boston's Black community, Baldwin was asked to serve on the executive committee of the local NAACP branch in its early days,[56] though again, Weiler suggests that Baldwin would have been too busy running her school to take a very active role.

Baldwin spent a number of summers leading teacher-training classes at two historically Black colleges, the Hampton Institute, now Hampton University, in Virginia and the Institute for Colored Youth, now Cheney University, in Pennsylvania.[57] (A women's dormitory at Howard University in Washington, DC, was named in her honor in 1950.)

One of the many educators whom Baldwin inspired was Charlotte Hawkins Brown, founder of a school for young African American men and women in Sedalia, North Carolina. Baldwin's achievements as an educator "not only gave Brown confidence but also stamped on whites an enduring impression of the intellectual abilities of African Americans. This recognition inspired Brown, who kept in contact with Baldwin and relied on her for advice," according to Brown's biographers.[58] Baldwin was one of many in the Boston area who supported Brown's school, reading the poems of Paul Lawrence Dunbar at a 1921 school fundraiser held in Cambridge.[59]

In 1920, Baldwin was named the first president of the League of Women for Community Service, which promoted the welfare of Black war veterans and African American students in the Boston area.[60] Baldwin was known for her optimism as well as her reformist fervor, and several commentators noted the lack of bitterness or recrimination in her manner and speeches.

Few records exist of Baldwin's views on the major issues of the day: her speeches were not transcribed, nor were her personal papers preserved in a library or archive. (Weiler holds out hope that Baldwin's extensive correspondence will eventually turn up in someone's attic.)

In 1900, Baldwin published a short essay called "The Changing Ideal of Progress" in the *Southern Workman* in which she contrasted the drive for personal improvement with the desire for societal progress, wondering whether one came at the expense of the other. She noted that "Quietly, but surely, there is growing a social consciousness. All the activities of life are being profoundly influenced by the deepening sense of the oneness of the human race. Ever stronger sets the tide of feeling against isolation, against segregation, of lives or of interests."[61] At a time when racial animosity was still widespread in the country, Baldwin expressed optimism that the opportunities available to Black Americans would improve with time.

The one essay notwithstanding, she largely kept her views on racial harmony out of the public eye. "She's very cautious in public life," Weiler says. That was part of how she managed to be successful in two worlds. "She was a very, very respected figure in the Black political community," while also winning widespread admiration among Cambridge's largely white political and educational leadership. An article in the *Cambridge Chronicle* in 1903 congratulates Cambridge on its history of promoting Black leaders—citing, among others, a city councilor, a fire chief, and the only female, Baldwin.[62] "The plucky woman has succeeded by talent and capacity, and her color seems to have cut no figure at all in the matter," the article reads.

On January 9, 1922, Baldwin collapsed at the Copley Plaza Hotel in Boston, where she had gone to deliver a speech at a fundraising event.[63] She died before arriving at the hospital. The next day at the Agassiz School, according to Cambridge author Pauline Elizabeth Hopkins, "the weeping heartbroken teachers could hardly control themselves enough to carry on the school and the children were full of awe and sadness."[64]

W. E. B. DU BOIS: PURE AMERICAN GENIUS

W. E. B. Du Bois was one of the foremost American intellectuals of the first half of the twentieth century. He is now acknowledged as a pioneering sociologist, conducting and publishing pathbreaking research on the lives of Black Americans. "He was a pure American genius," says Zine Magubane, a professor at Boston College.[65] Over a career spanning seven decades, Du Bois wrote seventeen books, including five novels; founded and edited four journals; and was both a scholar and a political organizer. "More than that," historian Thomas Holt says, "he reshaped how the experience of America and African America could be understood; he made us know both the complexity of who Black Americans have been and are, and why it matters."[66]

In 1905, Du Bois was instrumental in founding the Niagara Movement, an early civil rights organization and, in 1909, the National Association for the Advancement of Colored People (NAACP). By his example and in his writing, he advanced the idea that people of African descent are the intellectual equals of European Americans.

Despite all his accomplishments, Magubane, a sociologist who received her doctoral degree at Harvard in the 1990s, says she would never have known about Du Bois and his importance to her field if her father hadn't been a fan. In her years attending the same school where Du Bois received his undergraduate, master's, and doctoral degrees, she "never read a word of Du Bois. . . . I never encountered him in any intellectual context," she says. "As a sociologist trained at Harvard, the man's name was never uttered."

In 1903, Du Bois, the first African American to receive a PhD from Harvard, wrote *The Souls of Black Folk* about the existential challenge of being Black in America. Du Bois's book sold almost ten thousand copies in its first five years.[67] It begins: "Herein lie buried many things which if read with patience may show the strange meaning of being Black here at the dawning of the Twentieth Century. This meaning is not without interest to you, Gentle Reader; for the problem of the Twentieth Century is the problem of the color line."[68] Du Bois writes in *The Souls of Black Folk* that to be African American is to have one's identity defined by the prejudices and animosities of others. Articulating the aspirations of other Black Americans and arguably his own, Du Bois writes, "He simply wishes to make it possible for a man to be both Negro and an American, without being cursed and spit upon by his fellows, without having the doors of Opportunity closed roughly in his face."

Magubane says there are several reasons Du Bois was not given his due when she was a student. One is racism. Because Du Bois was Black, in the early twentieth century he was not accepted into the pantheon of modern sociologists like Émile Durkheim, Max Weber, and Karl Marx, she says. Another reason: in the years after World War II, Du Bois embraced Communism, convinced that capitalism would never properly serve the interests of African Americans. During the ideologically driven Cold War with the Soviet Union, Du Bois's admiration for Communism was considered an unforgivable sin. A third reason: earlier in his career, Du Bois had fought publicly with Booker T.

Undated portrait of W. E. B. Du Bois. *Source:* George Grantham Bain Collection, Prints and Photographs Division, Library of Congress, Washington, DC.

Washington, a powerful civil rights leader who argued for incremental improvements in Black rights while Du Bois insisted on nothing less than full equality. For a time, Magubane says, Du Bois "was persona non grata—even in the African American community." But he was nonetheless one of its great leaders.

Born in 1868, William Edward Burghardt Du Bois was raised in the small western Massachusetts town of Great Barrington, where his intellectual promise was first recognized. Part of a Black community that comprised perhaps 1 percent of the town's population, Du Bois never experienced the type of violent racism and segregation he later saw in the American South. Following his graduation from high school in 1885, Du Bois's Great Barrington neighbors—Black and white—paid for him to attend Fisk University, a historically Black college in Nashville, Tennessee. Three years later, degree in hand, he moved to Cambridge, where he received a scholarship to study at Harvard.

In his posthumously published autobiography, Du Bois recalled looking for a place to live in Cambridge because he would not be welcomed in the Harvard dormitories: "I tried to find a colored home, and finally at 20 Flagg Street, I came upon the neat home of a colored woman from Nova Scotia, a descendant of those Black Jamaican Maroons whom Britain deported after solemnly promising them peace if they would surrender. For a very reasonable sum, I rented the second story front room and for four years this was my home."[69] The house still stands, continuing in private ownership.

Entering Harvard, Du Bois was clearly a gifted writer and orator and confident in his skills, Magubane says. In an essay for a Harvard English class, he wrote, "I believe, foolishly perhaps, but sincerely, that I have something to say to the world, and I have taken English 12 in order to say it well."[70]

Du Bois was, above all, a serious student, though he believed that many of his peers were not: "Harvard of this day was a great opportunity for a young man and a young American Negro and I realized it. I formed habits of work rather different from those of most of the other students."[71] Du Bois was encouraged and inspired by several of his Harvard professors, including philosopher and psychologist William James, in whose home he was a regular guest, and historian Albert Bushnell Hart. "It was James with his pragmatism and Albert Bushnell Hart with his research method, that turned me back from the lovely but sterile land of philosophic speculation, to the social sciences as the field for gathering and interpreting that body of fact which would apply to my program for the Negro," Du Bois writes in his autobiography.

Du Bois describes himself as having been *in* Harvard but not *of* Harvard. He notes: "I was happy at Harvard, but for unusual reasons. One of these circumstances was my acceptance of racial segregation. Had I gone from Great Barrington high school

The Flagg Street home of W. E. B. Du Bois when he was a student at Harvard University, 2020. *Source:* Photo by the authors.

On Justice.

"Though Justice against Fate complain,
"And plead the ancient Rights in vain,
"But those do hold or break,
"As men are strong or weak."
                                    Marvell.

Judge Many with his big hands crowded far down in his pockets was rubbing his slippered feet thoughtfully. "I don't like," he said, "to consider Justice as something compulsory, as a law with a penalty. You can't say that the strong man owes the weaker a debt which he ought to pay. It is rather the beauty of strength that without compulsion, without sewed process for debt, it will help the helpless; and this is Justice. Justinian, (and Justinian was the good Judge's sole authority) gave but a shallow definition when he said 'Justitia est constans et perpetua — perpetuum — er — the broader, deeper, justice is not merely giving a man his own, it is giving him your own, in order that he may better

First page of an English 12 essay on the subject of justice written by W. E. B. Du Bois while a student at Harvard University. *Source:* David Graham Du Bois Trust, Special Collections and University Archives, University of Massachusetts Amherst Libraries. Used with permission.

directly to Harvard, I would have sought companionship with my white fellows and been disappointed and embittered by a discovery of social limitations to which I had not been used. But I came by way of Fisk and the South and there I had accepted color caste and embraced eagerly the companionship of those of my own color."[72] He socialized with Boston's Black intelligentsia: "My friends and companions were taken mainly from the colored students of Harvard and neighboring institutions, and the colored folk of Boston and surrounding towns. With them I led a happy and inspiring life. There were among them many educated and well-to-do folk; many young people studying or planning to study; many charming young women. We met and ate, danced and argued and planned a new world."

In 1890, when Du Bois was completing his bachelor's degree, he was invited to speak at Harvard's graduation ceremonies. The subject he chose for his speech was Jefferson Davis. "I chose it with deliberate intent of facing Harvard and the nation with a discussion of slavery as illustrated in the person of the president of the Confederate States of America. Naturally, my effort made a sensation. I said, among other things: 'I wish to consider not the man, but the type of civilization which his life represented: its foundation is the idea of the strong man—individualism coupled with the rule of might.'" Du Bois calls Jefferson Davis "the peculiar champion of a people fighting to be free in order that another people should not be free."[73] "Commencement came and standing before governor, president, and grave gowned men, I told them certain truths, waving my arms and breathing fast. They applauded with what may have seemed to many as uncalled-for fervor, but I walked home on pink clouds of glory!"

During his time at Harvard, Du Bois honed his thoughts on the subject of race in America and his skills as a philosopher, historian, sociologist, and writer. His 1895 doctoral dissertation, published as *The Suppression of the African Slave Trade to the United States of America, 1638–1870*, laid the groundwork for his later research and writing.

Magubane says Du Bois extended the ideas of social scientists like William James and employed them in a specifically racial context. James would ask, "What am I doing in this world, who am I?" she says, while Du Bois "would say every person has this struggle—but what do you do when you're having this struggle in the context of being forcibly identified by a larger society?" While to whites, he was identified solely as a Black man, Du Bois saw himself as a son of Great Barrington, with ancestry from Africa, Haiti, and the French Huguenot diaspora. "He looked at all the social processes that everyone else was studying and added another dimension," Magubane says.

Following completion of his studies at Harvard, Du Bois began his prolific academic and editing career. He taught at Wilberforce University in Ohio for two years. Then accepting a one-year research position at the University of Pennsylvania, he undertook a sociological study of Black people living in Philadelphia (published as *The Philadelphia Negro* in 1899), before settling at Atlanta University in Georgia, where he taught from 1897 to 1910. From 1910 to 1934, Du Bois edited the NAACP's publication *The Crisis*.

In his later years, Du Bois embraced socialism and then Communism as an alternative to the capitalist system that he saw as fostering oppression. In 1961, Du Bois became a citizen of newly independent Ghana, where he died on August 27, 1963. It is poignantly ironic that he died on the eve of the 1963 March on Washington at which Martin Luther King Jr. delivered his "I Have a Dream" speech, a turning point in the American civil rights movement.

Slowly, the broader field of sociology has come to recognize the contributions that Du Bois made to it. Magubane teaches a class at Boston College in "The Sociology of W. E. B. Du Bois." In October 2018, Harvard's sociology department hosted a three-day symposium to honor Du Bois on what would have been his 150th birthday.

MARRIAGE EQUALITY

On the afternoon of May 16, 2004, people began to gather on the terraced lawn of Cambridge's City Hall. The imposing Richardsonian Romanesque building loomed over them, but the atmosphere was festive. The crowd included excited and anxious couples, some toting young children, as well as a throng of well-wishers and journalists from all over the world. The mayor climbed a stepladder and asked people to form lines.

Inside City Hall, banisters were decorated with white tulle. A wedding cake and cups of cider awaited those who had come to receive application cards for marriage certificates.[74] Shortly after midnight, the city clerk's office opened, and Cambridge residents Marcia Hams and Susan Shepherd did something unprecedented in the history of the United States: they filed an application for a marriage license, becoming the first same-sex couple in America to legally do so.[75]

Kate and Nima Eshghi came to Cambridge City Hall planning to support two sets of friends who wanted marriage certificates.[76] They had two small children—ages two and five—so they weren't sure how the evening would go. But by the time they reached the inside of City Hall, they knew they wanted to stay, and they received the eleventh marriage license application issued that night. Emerging from the building several hours after entering, the couple remembers being astounded by the size of the crowd and the cheers that buoyed them as they walked down the steps.

That night, 270 couples applied for marriage certificates—ninety male couples, 174 female ones, and six heterosexual pairs—according to the *Boston Globe*.[77] "It was

Inside Cambridge City Hall as couples wait in line to .complete applications for marriage licenses, May 17, 2004. White bunting on the stair rails added to a celebratory atmosphere. *Source:* © 2004 Marilyn Humphries.

all-hands on deck for city workers, who showed up in droves to help move the process along smoothly."

Eshghi says that the climate that had developed in Massachusetts around marriage equality had left her feeling guarded. The whole previous year "was people arguing about the validity of our relationships, the validity of our family, what will happen to our children. There were a lot of negative and frankly insulting aspects of the debate." For Eshghi, the way Cambridge treated couples transformed the climate from a political battle into a celebration. "They really flipped the script that night," Eshghi says. "They made every couple feel special. That's how you're supposed to feel when you're getting married. We'd never experienced that before."

In 2001, GLAD, a Boston-based human rights advocacy group, had filed a legal case in Massachusetts, *Goodridge v. Department of Public Health*, on behalf of seven same-sex couples seeking the legal right to marry.[78] In November 2003, the Supreme Judicial Court of Massachusetts determined that the Massachusetts constitution did not allow the state to deny the benefits of marriage to couples of the same gender. Margaret H.

Local officials (including State Senator Jarrett Barrios, clapping, and city council member Denise Simmons, facing the camera) join in the celebrations at Cambridge City Hall, May 16 and 17, 2004. The event included speeches, music, wedding cake, and champagne. *Source:* © 2004 Marilyn Humphries.

Marshall, chief justice of the Massachusetts court, penned the majority opinion in the case. She wrote:

> Marriage is a vital social institution. The exclusive commitment of two individuals to each other nurtures love and mutual support; it brings stability to our society. For those who choose to marry, and for their children, marriage provides an abundance of legal, financial, and social benefits. In return it imposes weighty legal, financial, and social obligations. The question before us is whether, consistent with the Massachusetts Constitution, the Commonwealth may deny the protections, benefits, and obligations conferred by civil marriage to two individuals of the same sex who wish to marry. We conclude that it may not. The Massachusetts Constitution affirms the dignity and equality of all individuals. It forbids the creation of second-class citizens.[79]

Marriage license applicants Alex Fennell and Sasha Hartman swear an oath before Paula Crane, executive secretary to the mayor, in Cambridge City Hall, May 17, 2004. *Source:* © 2004 Marilyn Humphries.

The court stayed its decision for 180 days to allow the state legislature an opportunity to enact clarifying legislation, which it failed to do. As a result, city clerks in the state were free to issue marriage licenses to same-sex couples beginning on May 17, 2004. Alone in the state, the City of Cambridge Clerk's Office opened at midnight, welcoming people to apply for marriage licenses. In weddings that spring and summer, many couples included readings from Marshall's opinion in their ceremonies.

A few months earlier, in February 2004, the newly elected mayor of San Francisco, California, Gavin Newsom, had begun issuing marriage certificates to same-sex couples, but the legal status of those licenses was unclear, and they were later invalidated by the California Supreme Court.[80] California didn't enact marriage equality until 2008, a full four years after Massachusetts. In 2000, Vermont had authorized same-sex civil unions but not marriage.[81] As a result, Massachusetts became the first state to legalize nuptials between two men or two women.

Couples like the Eshghis remained anxious about whether the Massachusetts ruling would hold. Although they were among the earliest to complete their application, returning to Cambridge City Hall the next day to get the marriage license, the Eshghis wanted to wait until late summer to marry so their families could be with them. Kate Eshghi remembers carrying the license with her at all times, in case the legal status of their planned union suddenly changed and they needed to have a "quickie" wedding.

In Massachusetts, couples normally must wait three days after applying for licenses before they can be married, though a judge can waive the provision. Some same-sex couples sought waivers while others simply waited. In Cambridge, shortly after 9 a.m. on May 17, Tanya McCloskey and Marcia Kadish of nearby Malden took their vows, becoming the first legally wedded same-sex couple in America.[82] For the rest of that day, the steps and lawn of Cambridge City Hall were dotted with impromptu weddings and cheering crowds.

Couples who sought those early marriage licenses expressed a variety of motivations. For many, the main objective was simply societal recognition that they were a couple. Others wanted to establish legally recognized relationships for child custody or inheritance purposes. For still others, the medical visitation rights afforded an ailing spouse were important.

At the time, McCloskey and Kadish said they were simply making legal what had already been true for twenty years.[83] "We felt we were married already," Kadish told National Public Radio's *Morning Edition*. Unfortunately, the couple later became grateful for their marriage for a different reason. In 2015, McCloskey was diagnosed with endometrial cancer. Kadish told NPR that being a spouse rather than a partner meant she didn't have to worry about being allowed to support her wife during her medical care. "There was never a time that I couldn't see her in the hospital," Kadish said. "I pretty much didn't leave her side for almost a year. And we were respected. It was a beautiful thing, the support."

Despite threats that the state legislature or a higher court would overturn the Massachusetts marriage ruling, it remained in force. In 2015, the US Supreme Court's decision in the *Obergefell v. Hodges* case legalized marriage equality nationally. More than 32,000 same-sex couples were married in Massachusetts between 2004 and 2016, State House News Service reports.[84] What had once seemed inconceivable to many gay and lesbian couples became routine. The Commonwealth of Massachusetts does not track the number of same-sex divorces, but some research indicates a similar rate of divorce, about 2 percent per year, among same-sex couples as heterosexuals.[85] Julie Goodridge and Hillary Goodridge, the pair at the center of the state's 2003 court ruling, filed for divorce in 2009.[86]

Marcia Kadish and Tanya McClosky cheered by onlookers outside Cambridge City Hall after completing an application for a marriage license, May 17, 2004. *Source:* © 2004 Marilyn Humphries.

Although the specter of a more conservative US Supreme Court continues to temper the confidence of same-sex marriage advocates, Kate and Nima Eshghi have now celebrated more than three decades together. They were living in Somerville in 2004, but they made a point of holding their wedding ceremony in Cambridge. "There was this shared sense of joy," Kate remembers about that time. Nima quickly chimes in: "If you already didn't love Cambridge—we had always loved Cambridge—but we left [that day] feeling: Cambridge is so amazing. Look at this dynamic they created. We knew instantaneously we were going to get married in Cambridge."

INDUSTRIAL CAMBRIDGE

Rainer Weiss arrived at MIT for the first time in the fall of 1950, climbing the steps from the Red Line subway stop at Kendall Square. The future physicist and Nobel laureate was immediately struck. "It was the goddamned smelliest place that I ever walked into—and I come from New York," he later remembered. The stench was the result of emissions from the Lever Brothers soap factory, which was rendering animal fat a few blocks away, combined with those from a nearby pickle manufacturer, a rubber goods maker, and a chocolate factory or two.

Passing through Cambridge today and seeing the modern office towers, academic buildings, and tree-lined residential streets, it's hard to believe that the city was once an important industrial hub. But between 1860 and 1960, more than two hundred factories made a wide range of products, perfuming the air along the Charles River waterfront and up into North Cambridge. An exhaustive 1930 survey titled *The History of Massachusetts Industries: Their Inception, Growth and Success* devoted more than a hundred pages to descriptions of manufacturing enterprises in Cambridge.[1] The author noted that, by 1930, Cambridge had "experienced a metamorphosis that characterizes it as much of an industrial boom town as Akron, Ohio, or Detroit, Michigan."

Cambridge's industries were built on local assets: the ice that formed naturally in Fresh Pond each winter; the inexpensive reclaimed land along the Charles River in East Cambridge; a pool of available labor, often recent immigrants; and easy access to the financiers, patent attorneys, and stevedores just across the river in Boston.

Several of the stories featured in this chapter, such as the invention of the **sewing machine** and the beginnings of the **telephone**, focus on businesses that trace their origins to Cambridge even if they didn't end up among the city's major employers. Other stories highlight local industries that were both mainstays of the city's economy and truly innovative in their product development and marketing, such as the **international ice trade**, Carter's Ink Company (inventors of the **yellow highlighter**), and **Polaroid** (which introduced instant photography). Other manufacturing activities in Cambridge, like **candy making**, brought their products (such as the Charleston Chew and Junior Mints) to a national market though savvy marketing. Though

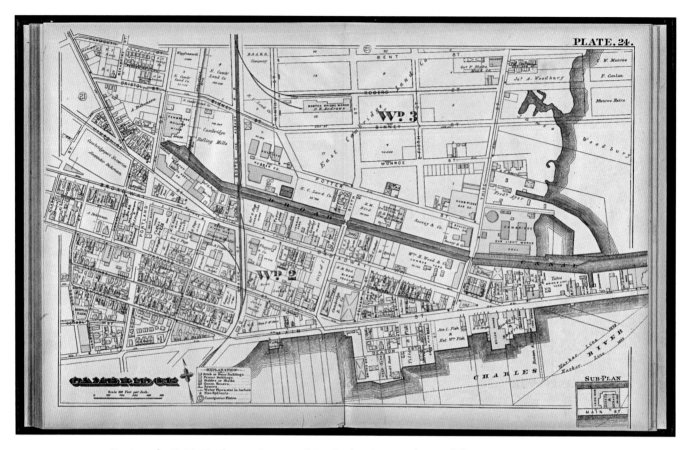

Portion of a G. M. Hopkins & Co. map of Cambridge showing the Kendall Square area, 1886. Note that the water line of the Charles River extended much closer to Main Street than it does today. *Source:* Norman B. Leventhal Map Center, Boston Public Library.

manufacturing activity tailed off in Cambridge after World War II, many of the local industries left behind sturdily constructed buildings that would prove useful in Cambridge's high-technology renaissance.

One of Cambridge's earliest large industries was glassmaking, which drew immigrants to East Cambridge, where they made decorative glassware. The New England Glass Company was founded in Cambridge in 1818, benefiting from new demand in the aftermath of the War of 1812. By 1851, New England Glass employed a staff of 450 in thirteen buildings, with annual revenues reaching $500,000. In 1888, the company, then run by Edward Libbey, relocated to Toledo, Ohio, to take advantage of lower fuel and labor costs. Eventually renamed Libbey-Owens-Ford, the company was for a time the largest glass company in the world.

The Chas. Davenport Car Manufactory building (1841–1857) at 700 Main Street, Cambridge. The site would be the future home of the Walworth Company, where Alexander Graham Bell completed the first long-distance telephone call and where Polaroid founder Edwin Land had his office and laboratory. The Davenport company became known for its center-aisle passenger rail cars. *Source:* Courtesy of the Cambridge Historical Society.

The following is a sampling of the diverse products manufactured in Cambridge in 1930:

Bricks and clay pots   Beginning in the 1840s and continuing for more than a century, bricks and flowerpots were made from the extensive clay pits found in the Alewife Brook area of North Cambridge. In the 1850s, bricks were produced in Cambridge at the rate of 187,000 a day. By World War II, the clay deposits had been nearly exhausted. The flowerpot factory closed in 1934, and local brick production ended in 1956. Many of those former pits are now parkland, including the city's largest, Danehy Park.

Processed meats   The John P. Squire Company was a large meat-packing enterprise in East Cambridge from 1855 to the 1950s. By 1930, the firm slaughtered as many as six thousand hogs a day at its twenty-two-acre factory on Gore Street, employing more than a thousand workers. The factory, abandoned in the late 1950s, burned to the ground in 1963.

Steam boilers   The Kendall Boiler and Tank Company, namesake of Kendall Square, was established in 1860 to manufacture industrial steam boilers, used as a power source in many New England factories before high-voltage electrical distribution became common. In 1896, the company employed two hundred.

Custom furniture   The A. H. Davenport Company began manufacturing high-quality furniture in East Cambridge around 1880 and built custom furnishings for several prominent American architects, including H. H. Richardson. Davenport's products were chosen to furnish the White House during Theodore Roosevelt's presidency (1901–1909). In 1909, William Howard Taft's administration used Davenport pieces in the first Oval Office. In many parts of the country, a fancy sofa is still called a davenport, though most people are unaware of the word's connection to Cambridge. Decades later, the Davenport building served as headquarters for two tech companies: Zipcar and HubSpot.

Fire hoses and rubber products   The Boston Woven Hose and Rubber Company was established in Cambridge in 1884 and built a large manufacturing complex on Hampshire Street in Kendall Square. The firm manufactured hydraulic hoses, bicycle inner tubes, and other rubber goods. Its breakthrough product was the seamless fire hose. By 1930, its fifteen-acre plant included nineteen buildings and employed twelve hundred. But with an aging multistory plant, the company struggled to compete after World War II and in 1956 was acquired by a larger firm. The company discontinued production of rubber products in Cambridge in 1969. The building complex it abandoned later served a cluster of e-commerce start-ups, data analytics firms, and biotechnology companies.

Soap The Lever Brothers Company, a British manufacturer of soap products, established its US headquarters in Cambridge in 1898 by purchasing an existing soap factory. Today part of the Unilever conglomerate, Lever Brothers made its Lifebuoy soap, Lux soap flakes, and Rinso detergent in Kendall Square until the late 1950s, when the company left behind its aging facilities there. Before the invention of synthetic detergents, most soap was made from animal tallow, the fat removed from slaughtered cattle and pigs. The complex was redeveloped by MIT and a private developer in the 1960s into what is now Technology Square, a seven-building, 1.2-million-square foot business park employing as many as 2,500 people.

Insulated electrical cables The Simplex Wire and Cable Company on Franklin Street employed 750 people in the manufacture of insulated electrical wire and undersea transmission cables. In the 1950s, the firm's undersea cables spanned from Florida to Cuba and from Port Angeles, Washington, to Ketchikan, Alaska. Many cables were produced for the US military, including cables from Vietnam to Thailand. In 1969, the firm was sold, and production moved to Maine. MIT bought the extensive Simplex property between Sydney and Brookline streets, demolishing the plant and creating University Park, a mixed-use development with offices, laboratory space, and housing.

Automotive fasteners In 1912, the Carr Fastener Company began operations, making metal fasteners that were used to attach canvas to horse-drawn carriages and early automobiles. In 1929, the company became United-Carr and relocated to Binney Street. It diversified its product lines, which served the automotive and boat-building industries, by developing snaps and uniform buttons for garment makers and manufacturing specialty hardware for fastening many kinds of metal, wood, and fabric components. The company was sold in 1969 and discontinued operations sometime prior to 1995. The Binney Street building complex was converted to loft apartments.

The stories that follow feature some of the most innovative products of industrial Cambridge: the international ice business, the first sewing machine, the beginnings of the telephone, brands of candy that remain popular today, instant photography, and the yellow highlighter.

SLIPPERY SPECULATION: THE INTERNATIONAL ICE TRADE

Local lore has it that the international ice business started in Cambridge. The truth is a little more complicated.

In 1806, people thought it was hilarious when twenty-three-year-old Bostonian Frederic Tudor announced that he was going to ship ice from his father's farm in Saugus, Massachusetts, to countries in the Caribbean—places where no one had ever seen frozen water.[2] To be of any value, the ice would have to stay solid during a weeks-long sea journey. "No joke," the *Boston Gazette* newspaper wrote at the time. "A vessel with a cargo of 80 tons of Ice has cleared out from this port for Martinique. We hope this will not prove to be a slippery speculation."[3] Tudor himself thought the ice business was such a brilliant idea that he kept news of his venture out of the papers until the last minute—worried that competitors might soon appear.

In a way, both Tudor and his detractors were right. Some of the ice *did* stay frozen in the hold of the ship despite the long distance. But it didn't stay frozen long in the heat of the Caribbean sun. It would take many years before Tudor earned a regular return on his investment. Although the product was freely available in winter across New England, getting ice into a shippable form was a laborious process. Tudor needed to build double-walled ice houses near the Caribbean ports to prevent melting. He was forced to buy and staff his own ship because "merchants were not willing to charter their vessels to carry ice. The officers declined to insure and sailors were afraid to trust themselves with such a cargo," according to Tudor's brother-in-law Robert Gardiner.[4]

For a quarter century, from 1805 until the early 1830s, Tudor struggled to make a living from his venture. He ended up in debtor's prison twice. Ships transporting his ice sank. Occasional warm winters meant he had to harvest his stock in Maine instead of Massachusetts, adding to his costs.[5] He struggled to find competent agents in faraway places and to defend exclusive sales rights he had painstakingly negotiated.

Demand was low at first. So Tudor turned to marketing. At his direction, Tudor's agents encouraged barkeeps to mix cold drinks and sell them for the same price as those at room temperature.[6] "A man who has drank his drinks cold at the same expense for one week can never be presented with them warm again," Tudor wrote in a letter to his business agent in Martinique. He provided instructions on how to make ice cream. He also pitched ice as medicine for fighting the fevers that were common in tropical environments.

Tudor's financial (mis)fortunes began to turn in 1826 when he took on twenty-four-year-old Nathaniel Jarvis Wyeth as a foreman.[7] Wyeth's family owned a hotel on a bluff overlooking Cambridge's Fresh Pond, and it became a popular summer retreat for Bostonians.[8] Expected to attend Harvard, Wyeth had instead gone to work at his family's hotel. There he often harvested ice for hotel guests, selling the surplus to local retail merchants and Tudor, who had long relied on Fresh Pond for his ice supply.

ICE-CUTTING AT FRESH POND, CAMBRIDGE, MASS.

Engraving of ice harvesting on Fresh Pond, Cambridge. *Source:* Collection of the authors.

After years of painstakingly chopping his quarry from the frozen pond with axes and saws, in 1825 Wyeth perfected a more efficient way of carving up the ice. A horse-drawn cutter first scored the pond's surface into a giant checkerboard of standard-size blocks. Using a second plow, he carved deeper. A third device allowed Wyeth to plane the ice to a consistent thickness. Men with hand tools then carved the slabs and floated them to a nearby icehouse, where they were cut into pieces and carefully stored to make maximum use of space. As many as a dozen teams of horses and a hundred men might be needed to fill a single ice house at the peak of the winter carving season.

While ice harvesting was still laborious, Wyeth's inventions—variations on a standard farm plow and a carpenter's plane—enabled the mass production of ice.[9] The giant, regularly shaped blocks remained frozen longer, especially when Wyeth packed them in sawdust. Impressed by his ingenuity, Tudor hired Wyeth to manage the supply side of the business for $500 a year—a good deal for a young man eager to escape his family's hotel business.[10]

Frederic Tudor, engraving, ca. 1888. *Source:* From D. Hamilton Hurd, ed., *History of Essex County Massachusetts, with Biographical Sketches of Many of Its Pioneers and Prominent Men* (Philadelphia: J. W. Lewis & Co., 1888).

But he didn't last long. Wyeth quit the ice business in 1832 and led two expeditions to the Pacific Northwest, convinced he could make a fortune trading in furs and salmon. In 1834, he established Fort William, a trading post on the Columbia River, near what became Portland, Oregon.[11] Wyeth eventually returned to Boston and, after a falling out with his former boss, became Tudor's competitor.

Meanwhile, Tudor's business, buoyed by Wyeth's inventions, became profitable. Sales were strong in places like Charleston, South Carolina, and New Orleans, Louisiana, and later in port cities in India, where British families suffered in the heat and dreamed of the comforts of home, including iced drinks. In 1833, when Tudor's ice first arrived in Calcutta, India, after a journey of some 14,000 miles, the *Calcutta Courier* wrote that "the names of those who planned and have successfully carried through the adventure at their own cost, deserve to be handed down to posterity with the names of other benefactors of mankind," such as the first importers of the potato to Europe, the newspaper claimed.[12]

The industry formed by Tudor's vision and Wyeth's ingenuity "would within twenty years or so transform America into the first refrigerated society, and make New England a major exporter of ice to countries as far away as India," journalist Gavin Weightman wrote in 2003.[13]

So many businessmen began vying for the ice on Fresh Pond that Wyeth suggested they devise a fair way to divvy it up. In 1841, a committee led by Harvard law professor Simon Greenleaf proposed that ownership of the shoreline should determine rights to the ice in the pond—so whoever owned 30 percent of the shore was entitled to 30 percent of the ice.[14] This decision caused an explosion in property values along the pond and became a national model for resolving similar property disputes. Tudor, who had bought the 120-acre Fresh Pond Farm in 1838 for $130 an acre, turned down an offer a decade later for $2,000 an acre. Wyeth and Tudor were the largest landowners on Fresh Pond and secured access to the most ice. Tudor also created a seven-acre lake on low-lying areas of his land to expand his harvest. In 1837, Tudor's business on Fresh Pond employed 127 men, 105 horses, and one bull.[15] By 1847, Wyeth, Tudor, and their competitors were shipping some fifty thousand tons of ice a year.[16] Tudor's ice business had turned into an industry with him, "the Boston Ice King," at its helm.[17]

During the winter of 1846–1847, the essayist and philosopher Henry David Thoreau was so disturbed by Tudor's workers, who were carving ice from his beloved Walden Pond in Concord, that he wrote in a letter to his friend Ralph Waldo Emerson that

> [A] hundred Irishmen, with Yankee overseers, came from Cambridge every day to get out the ice. They divided it into cakes by methods too well known to require description, and these, being sledded to the shore were rapidly hauled off on to an ice platform, and raised by grappling irons and block and tackle, worked by horses, onto a stack, as surely as so many barrels of flour, and there placed evenly side by side, and row upon row, as they formed the solid base of an obelisk designed to pierce the clouds.[18]

Wyeth died in 1856 at the age of fifty-four a wealthy man and was buried in Mount Auburn Cemetery, just a short walk from where he lived most of his life. After establishing his own substantial fortune, Tudor died eight years later at age eighty.

Commercially sold ice became a huge business during the nineteenth century, transforming food production and making everyday items out of previously rare luxuries like ice cream. By the beginning of World War I, production shifted to factory-made ice, and the ice industry no longer depended on nature for its supply, although people in rural areas of the country continued to cut winter ice for ice boxes until rural electrification became commonplace in the 1940s and 1950s.[19]

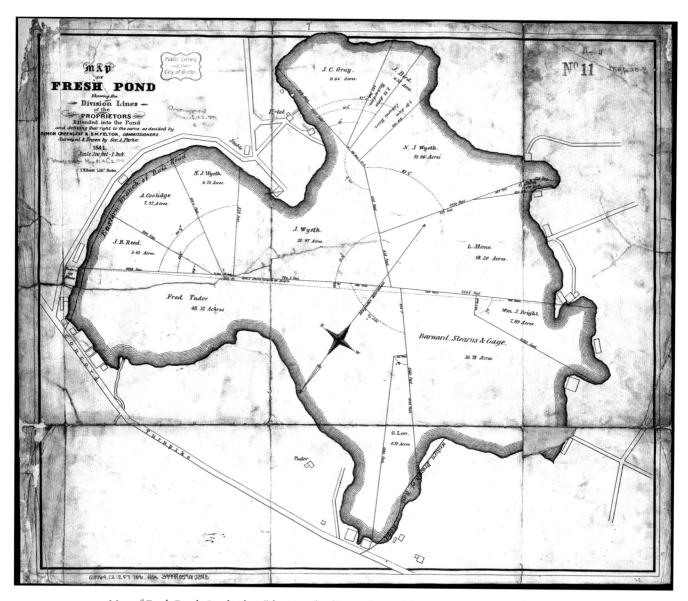

Map of Fresh Pond, Cambridge, "showing the division lines of the proprietors extended into the pond and defining their right to the same as decided by Simon Greenleaf & S. M. Felton, commissioners," 1841. *Source:* Norman Leventhal Map Collection, Boston Public Library.

NATHANIEL J. WYETH.

Portrait of Nathaniel Jarvis Wyeth from *Harper's New Monthly Magazine*, November 1892. *Source:* Wikimedia Commons.

Fresh Pond is now a reservoir, a holding area for Cambridge's water supply. It is fenced off and surrounded by walking paths. The pond remains slushy even in midwinter; it's hard to imagine that it was ever consistently frozen enough to ride across with horse-drawn plows. The hotels and ice houses were demolished long ago.

In their day, Tudor and Wyeth, transformed two essentially free products (natural ice and sawdust), transported them inside ships in need of ballast, built an industry, and created a pair of fortunes. "Above all," Weightman wrote, the ice trade "furthered the reputation of New England merchants as ingenious and benevolent entrepreneurs."

ELIAS HOWE'S SEWING MACHINE

Elias Howe, a journeyman mechanic, was already well acquainted with failure by age twenty. He wasn't physically robust enough to be a farmer. Work as a factory hand in the textile mills in Lowell, Massachusetts, had proven unsteady. So in 1839, Howe found himself living in Cambridge and earning a meager living in the Boston workshop of a man named Daniel Davis.[20]

One day in 1842, a wealthy man came into Davis's workshop. He had enlisted the help of an inventor in an unsuccessful effort to design a knitting machine and hoped Davis could achieve what his inventor had not. Asa Davis, brother of Daniel, chided the man.[21] "What are you bothering yourselves with a knitting machine for? Why don't you make a sewing machine?" he asked. "I wish I could," the man answered, "but it can't be done." "O yes it can," Davis responded boldly. "I can make a sewing machine myself." "Well, you do it Davis and I'll insure you an independent fortune," the man replied.[22]

The conversation, as Elias Howe recounted it to a biographer three decades later (though it is perhaps apocryphal), made an impression on him. It triggered "a habit of reflecting upon the art of sewing, watching the process as performed by hand and wondering whether it was within the compass of the mechanic arts to do it by machinery."[23] He thought the human act of sewing was a waste of energy and imagined "it was the very work for a machine to do."[24]

By 1840, Howe had married. The couple soon had three children, stretching his $9-a-week salary. He often complained of exhaustion, skipping dinner to crawl into bed, eager "to lie in bed forever and ever."[25] His wife took in sewing work to help make ends meet.[26]

Desperate to improve his lot in life, in 1843 Howe began to experiment with possible designs for a sewing machine.[27] He spent months working on a machine that might mimic his wife's exact movements as she sewed. He could never get it to work.

In 1844, he was struck with an idea: instead of trying to copy the actions of a seamstress, what if the machine sewed differently? Howe soon imagined using two threads and forming a stitch with the help of a shuttle and a curved needle. He made a rough model from wood and wire that October, proving to himself that his invention could work.

Howe quit the Davis shop to work full-time perfecting a model of his sewing machine. He was supported by his father, who had moved to Cambridge to make hats, using a machine invented by his brother Tyler.[28] Elias Howe Sr. ran a hat factory at 740

Main Street. But the hat-making machinery was destroyed in a fire in November 1844, robbing the family of its main source of income.

A boyhood friend, George Fisher, stepped in to support Howe's efforts.[29] By the end of 1844, Fisher, a Cambridge coal and wood merchant, was housing the Howes and had provided $500 for the materials and tools needed to build a working model of the sewing machine.[30] In return, Howe promised Fisher half his future earnings. "I believe," Fisher testified in a later patent suit, "I was the only one of his neighbors and friends in Cambridge that had any confidence in the success of the invention. He was generally looked upon as very visionary in undertaking anything of the kind, and I was thought very foolish in assisting him."[31]

181 Main Street, Cambridge, 1899. Elias Howe lived in the building, which was owned by his friend George Fisher, during his development of the sewing machine. The building was demolished in the early twentieth century. *Source:* Main Street Collection, Cambridge Historical Commission.

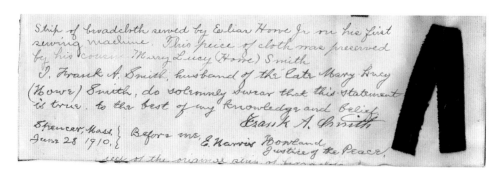

A piece of fabric sewn by Elias Howe on his first sewing machine. "The strip was preserved by his cousin, Mary Lucy Howe. Later, it was notarized and signed by her husband, Frank A. Smith, on June 28, 1910, and presented to the Historical Society of Spencer, Massachusetts." *Source:* Spencer Historical Museum Collections, Richard Sudgen Library, Spencer, MA.

In April 1845, Howe mechanically sewed his first seam. By July, he had sewn two suits—one for himself and another for Fisher. Seeking to stir up interest in his invention, Howe demonstrated his machine to several tailors, who discouraged him, fearing he would put them out of work. Howe took the machine to the Quincy Hall Clothing Manufactory in Boston, where every day for two weeks he offered to sew anything the tailors could offer. The machine sewed at 250 stitches a minute and could make ornamental stitches.[32] Howe challenged five of the fastest seamstresses to a race: they had to sew one seam apiece while he stitched five. Howe finished first, and the man who held the stopwatch, himself a tailor, swore in a statement that "the work done on the machine was the neatest and strongest."

Still tailors resisted—either on principle, out of fear, or because of the high price of the machine, which had cost Howe $300 to build.[33]

Howe set out to patent his sewing machine, which required him to build a working model to submit to the US Patent Office. In late summer 1846, at Fisher's expense, the two men exhibited the machine in Washington, and on September 10, they were awarded US Patent 4,750.[34]

Fisher grew tired of supporting a project that had cost him $2,000 and still held no promise of financial return.[35] "I had lost confidence," he later testified, "in the machine's ever paying anything." In October, Howe set off for England with one of the machines.[36] He negotiated a sale to William Thomas, who employed five thousand people sewing corsets, umbrellas, bags, and shoes. Thomas paid $250 for the machine and the right to hold the patent in England. Early the next year, Howe, his wife, children, and brother returned to England at Thomas's expense.[37] For eight months, Howe

Elias Howe's 1846 patent model for his sewing machine. *Source:* Harold Dorwin, National Museum of American History, Smithsonian Institution, Washington, DC. Used by permission.

worked for the factory owner, adjusting his machine so it could stitch around the stays that gave shape to the umbrellas and corsets. Money was tight, and Howe sent his wife and children back to Cambridge to live with his father.[38] To pay for his own passage home, he pawned his original machine and his patent letters. He arrived in New York in April 1849 without enough money to complete the journey back to Cambridge. He quickly got work in a machine shop in New York to earn his way home but hadn't yet earned enough when he got word that his wife was dying of consumption. His father sent him $10 for the trip home, and he was by his wife's side when she died. He had to borrow a suit for the funeral. Then the ship carrying all the family's household goods back from England sank off Cape Cod.

Back in Cambridge, Howe was shocked to learn that sewing machines based on his patented design were being used across New England.[39] That summer, working again as a journeyman mechanic, he raised $100 and entrusted a friend who was heading to England to reclaim his machine and patent letters and send them home. Once the letters

In 1940, the US Post Office issued a postage stamp featuring Elias Howe as part of its Famous American Inventors series. *Source:* Photo by the authors.

were back in his possession in the fall of 1849, Howe contacted the firms he believed were infringing his patent rights, offering to license his technology for a fee. He was generally ignored.

Planning to pursue legal action against the infringing firms, Howe convinced a man named George Bliss to buy Fisher's share of the patent rights and pay for the lawsuits. As the cases wound their way slowly through the courts, Howe continued to manufacture sewing machines from a small workshop in New York.

Once again, Howe hit a roadblock. This one was named Isaac Merritt Singer. Singer had built his own sewing machine similar in design to Howe's, after seeing an early model in Boston. Singer was unwilling to concede that Howe had invented the first practical sewing machine. In 1854, a court in Boston upheld Howe's patent and charged Singer and the others with infringement. "There is no evidence in this case that leaves a shadow of doubt that, for all the benefits conferred upon the public by the introduction of a sewing machine the public are indebted to Elias Howe," a judge on the case ruled.[40]

The Elias Howe monument in Bridgeport, CT, 2020. *Source:* Courtesy of Richard Anderson.

Bliss died a short time later, and Howe bought out his share of the patent, becoming the sole owner of his own invention for the first time.

Howe achieved his goal of making an independent fortune. As an outcome of the 1854 court decision, manufacturers had to pay Howe $5 for every sewing machine sold in the United States and $1 for each one sold overseas, for a period of six years. After 1860, the year he received a seven-year patent extension, Howe received $1 for every machine sold. Sewing machines became such a sensation that in 1863 Howe's royalties were estimated at $4,000 a day.[41] By the time the patent expired four years later, Howe had earned close to $2 million. He died the same year, at age forty-eight, in Brooklyn.[42]

In 1940, Howe was featured on a five-cent US postage stamp, part of the Famous American Inventors series. A monument in his honor was erected in 1884 in Bridgeport, Connecticut, where he built a factory in the 1860s to produce his sewing machines. But as of this writing, no marker in Cambridge highlights Howe's achievement.

THE BELL TELEPHONE COMPANY AND GARDINER GREENE HUBBARD

Alexander Graham Bell is widely considered the inventor of the telephone, yet it is a less well-known man, Gardiner Greene Hubbard, a Cambridge lawyer and entrepreneur, who may deserve an equal measure of credit for Bell's success.

At 700 Main Street in Cambridge, a bronze wall plaque memorializes an event that shaped the future of modern telecommunications. "From this site on October 9, 1876," the plaque reads, "the first two-way long-distance conversation was carried on for three hours. From here in Cambridgeport, Thomas G. Watson spoke over a telegraph wire to Alexander Graham Bell at the office of the Walworth Mfg. Co., 69 Kilby Street, Boston, Mass." To test the accuracy of the transmission, both Bell and Watson wrote down the words that each had spoken and each had heard and later compared notes.[43]

As Bell recounted in a legal deposition in 1887,

> The results [of the experiment] were very satisfactory, and sustained conversation was, for the first time, carried on by electrical means between persons who were miles apart.[44]

Bell, a native of Scotland, emigrated to Canada in 1870 at the age of twenty-three along with his parents. In his family's Ontario home, Bell began conducting experiments with electricity and sound. Arriving in Boston two years later, Bell established a private practice as a speech teacher working with the deaf. Among his students was Mabel Hubbard, the fifteen-year-old daughter of attorney Gardiner Hubbard. Mabel

Alexander Graham Bell, ca. 1904. *Source:* Library of Congress

Bronze plaque at 700 Main Street, Cambridge, honoring Alexander Graham Bell and the first long-distance telephone conversation. *Source:* Photo by the authors.

had lost her hearing at the age of five from scarlet fever. Bell also began teaching speech at Boston University and continued his laboratory experimentation.

Hoping to cut back on his private teaching, in 1874 Bell approached Gardiner Hubbard, looking for a financial backer who could underwrite his research. Hubbard had a history as a venture capitalist and entrepreneur; he had organized the city of Cambridge's first water works, founded the Cambridge Gas Company, and would later invest in a street railway connecting Cambridge to Boston. Hubbard agreed to a three-way partnership with Bell and Thomas Sanders, the father of another of Bell's students.[45] Bell continued work at his makeshift laboratories in Boston and at his family home in Canada.

Bell's initial goal was to improve telegraph transmission with a device that would allow the simultaneous transmission of multiple messages across a single telegraph wire, using different frequencies. In the 1870s, the telegraph was big business, and it would be even more lucrative if inventors could figure out how to send more messages over the existing network of wires. It was this potential that attracted Bell's investors. As historian Christopher Beauchamp noted:

> These men plunged into a race to develop sound-based electrical signals, a contest which led unexpectedly to the transmission of human speech. What happened next—the commercialization of the telephone as a disruptive stand-alone technology, independent of the telegraph—was not inevitable. Nor was it the result of visionary choices by an individual inventor. Instead, the fate of the new device depended heavily on the character of the Bell enterprise: a high-tech start-up whose strategies were driven by investor relations and the exploitation of patents.[46]

With his experience as an attorney, Hubbard advised Bell to document and date every experiment and potential technical advance.[47] This advice proved crucial to Bell's success in patent infringement lawsuits in which he later became embroiled. On March 7, 1876, Bell's first telephone patent, number 174,465, was issued, though he had not yet conclusively produced a working telephone. A few days later, on March 10, Bell uttered the now famous words "Mr. Watson, come here, I want to see you" at his laboratory at 5 Exeter Place in downtown Boston, proving that sound could be transmitted electrically over a wire.[48] In June, he demonstrated, to much acclaim, the transmission of vocal sounds over an electrical wire at the Centennial Exhibition in Philadelphia.

A legal challenge to Bell's patents was filed in 1880 and decided the following year.[49] In that claim, the judge ruled that Bell had "discovered a new art—that of transmitting speech by electricity—and has a right to hold the broadest claim for it which can be permitted in any case."

If Bell had remained in rural Brantford, Ontario, he might have been hard-pressed to devise some of the components of his working telephone system. But in Boston, he had access to many resources. In addition to Hubbard's legal and business advice and financial support, Bell drew on the engineering know-how of MIT (then located in Boston's Back Bay) and the specialized fabrication expertise of Boston's machine shops. At MIT, Bell attended lectures on physics and electricity. In the Boston workshop of Charles Williams, Bell found laboratory space, an equipment supplier, and the mechanic Thomas Watson.[50]

In November 1876, Bell continued his experiments in long-distance telephony. This time, he made use of a telegraph line running between the Harvard College Observatory on Garden Street in Cambridge and the downtown Boston office of an electrical company. The observatory's line was used to transmit accurate time, astronomically derived, to Boston's clockmakers. Bell recounted that he was "in the habit of going out to Mr. Gardiner G. Hubbard's house in Cambridge very frequently in the evening." The line wasn't generally needed by the observatory at night, so Bell was able to arrange to use it for his evening experiments, a situation that continued for some time. "It was my custom, when some new modification had been made, to carry out a telephone to Cambridge, and test it, at night, on the line running to the laboratory, Mr. Watson being at the Boston end."[51]

There was another reason Bell was spending so much time in Cambridge: love. In the fall of 1875, Bell became engaged to Mabel Hubbard. When the two married on Hubbard's Brattle Street estate in July 1877, Mabel was nineteen, Bell a full decade older.[52]

Bell and his business partners formed the Bell Telephone Company in Boston on July 9, 1877. Hubbard, Bell, and Sanders each received 30 percent of its shares, with another 10 percent allocated to Thomas Watson. Bell assigned all but ten of his shares to his wife Mabel. She in turn gave her father power of attorney over her stake, placing Hubbard in control of the majority of the company's shares. He became the Bell Telephone Company's de facto president.

It was Hubbard who decided on a policy of leasing, rather than selling, telephone equipment to end users. This strategy, which Hubbard had learned from his legal work for a Boston-area shoe machinery company with a similar policy, would prove immensely profitable to the telephone company. As late as the 1970s, customers of American Telephone and Telegraph (AT&T), the Bell Company's corporate successor, still leased rather than owned their own phones.

Hubbard was instrumental in creating several interconnected corporate entities to leverage Bell's telephone patents, including the New England Telephone and Telegraph Company in 1878 and, to promote the company's technology in Europe, the International Bell Telephone Company the following year.

Gardiner Greene Hubbard and Gertrude Hubbard, parents of Mabel Hubbard and parents-in-law of Alexander Graham Bell, ca. 1890–1897. *Source:* Gilbert H. Grosvenor Collection of Photographs of the Alexander Graham Bell Family, Prints & Photographs Division, Library of Congress.

Letter from Alexander Graham Bell to Gardiner Greene Hubbard regarding his stake in the Bell Telephone companies in England and continental Europe, July 28, 1880. In the letter, which Bell wrote from Cambridge, he admitted his reliance on Hubbard's business acumen in managing his and Mabel's financial interests in the telephone business. *Source:* Library of Congress.

Gardiner Hubbard moved permanently to Washington, DC, in the 1880s and subdivided his Cambridge property into house lots. He commissioned some of Boston's finest architects to design stately homes on the parcels, ostensibly for sale to Harvard professors. The houses built between 1887 and 1894 along Hubbard Park Road and Mercer Circle represent Hubbard's remaining legacy in Cambridge.

In an 1897 obituary, the journal *Science* (which Hubbard had himself founded) noted: "Mr. Hubbard was not the discoverer of the laws of acoustics which are represented in the telephone; he was not the inventor of the telephone, but he was the entrepreneur who distributed the telephone among all men of the civilized world and made it a practical agency for social intercommunication."[53]

THE SWEET LIFE IN CAMBRIDGE: CANDY MANUFACTURING

On many days, the scent of chocolate and mint hangs in the air outside 810 Main Street in Cambridge. A small sign on the five-story building's facade reads simply "Cambridge Brands." Painted-over window panels reveal nothing of the activity inside. The owners regularly decline to give tours.[54] The lack of openness adds to the mystique of Cambridge's last remaining large-scale candy maker. The factory was built for the James O. Welch Company in 1927 and launched several familiar candies: Sugar Daddies (1926), Sugar Babies (1935), and Junior Mints (1949). As of 2020, all three were still made here. Reportedly, more than fifteen million Junior Mints are produced each day at the factory.[55]

Cambridge was once a major candy-making center, home to many well-known brands, including Charleston Chew, Squirrel Nut Zipper, Conversation Hearts (the Valentine's Day candy), and Necco wafers. In 1907, the *Cambridge Chronicle* reported that there were seventeen businesses in the local confectionery business, employing eight hundred workers and producing more than $1.6 million of goods annually.[56] Forty years later, the number of candy manufacturers had quadrupled.[57] Many of these were arrayed along "Candy Row"—the stretch of Main Street between Kendall Square and Central Square where Cambridge Brands is still located.

Cambridge's first foray into the candy industry began in 1820, when a Connecticut man named Isaac Lum moved here, setting up shop at 165 Broadway.[58] Lum taught the candy trade to teenager Robert Douglass, who later started in the candy business himself, "trundling his sugar from Boston in a wheelbarrow, and in like manner carrying his finished products to the larger city for sale."[59] The Revere Sugar Refining Company began sugar production in East Cambridge in 1871, producing candy's main ingredient and cementing the role of candy in the city's industrial base. "From a comparatively

The manufacturing plant of Cambridge Brands on Main Street in Cambridge, outside of which the sweet smells of chocolate, mint, and caramelized sugar often hang in the air, 2020. *Source:* Photo by the authors.

small beginning, this industry has grown to be one of the most important in the city and is now so great as to make Cambridge conspicuous among the candy producing cities of the United States," the *Cambridge Chronicle* noted in 1907. "The confections manufactured in the factories of this city are known throughout the whole country and many of them are famous as the best products of their kind."[60]

During World War I, US soldiers received individually wrapped candy bars in their rations as a portable source of concentrated calories.[61] Returning home after the war, soldiers continued to consume candy. The 1920s became something of a golden age for the candy bar in the United States. Thousands of new confectionery products—mostly combinations of chocolate, caramel, nougat, sugar, and nuts—were introduced between the two world wars. At a time when malnutrition was a larger problem than obesity, candy manufacturers touted the "food value" of their products, suggesting that candy was an inexpensive alternative to meat, milk products, and other more conventional comestibles.[62]

Vintage Charleston Chew box. *Source:* Collection of the authors.

By the late 1920s, candy making was the second-largest industry in Cambridge, producing chocolates, hard candies, peanut candies, and walnut fudge.[63] But candy is a fickle business. During economic downturns, people buy less.[64] And people's tastes change. The candy industry in Cambridge relied on chocolate, sugar, and other raw materials transported from great distances, adding costs to the end product. After World War II, as newly built highways, truck transportation, lower labor costs, and inexpensive land made manufacturing more competitive outside of city centers, the candy industry in Cambridge began a slow decline.

Ice cream maker Gus Rancatore opened his first Toscanini's store near Central Square in Cambridge in 1981, "when the neighborhood was sweet with candy making and tank trucks, and rail cars heavy with liquid sugar blocked streets outside of working candy factories."[65] Rancatore's store gave him a front-row seat to watch the industry's demise. "If you had a sense of urban history," he says, "you could see that this was a nineteenth-century business that was giving way to something else."[66]

Some of the sweets produced in Cambridge became household names, while others have faded from memory. Here are a few examples.

Toasted Marsh Mallow Muffins

The George Close Company was established in East Cambridge in 1872 and built a new factory at the corner of Windsor Street and Broadway in 1879, expanding it in 1911.[67] The Close company made penny candy, as many as 150 varieties, and sold in bulk to wholesalers and retail stores. Its product line included lemon drops, butter balls, and Christmas candy, sold mostly in New England. By 1907, the company was producing six tons of candy a day, including a treat called Close's Toasted Marsh Mallow Muffins, which was shipped across the eastern United States.[68] The firm employed two hundred people in 1930, mostly young women, about half of whom were the children of immigrants.[69] The company failed in the Great Depression. The baseball cards that the company released with some of its candies in the 1910s, featuring players like Cy Young and Ty Cobb, remain collector's items.

Squirrel Nut Zippers

In 1903, the Squirrel Brand Salted Nut Co. moved to Cambridge from Boston. Former-worker-turned-owner Perley G. Gerrish traveled by horse and wagon to bring his mixed nut varieties to stores across the region. Squirrel Brand built a new factory on Boardman Street in 1914. Polar Explorer Admiral Richard Byrd carried Squirrel Brand nuts with him to the South Pole, according to the company's history.[70] During World War II, Squirrel Brand nuts were shipped across the world, with one soldier in the Philippines writing back to the company that his comrades' Christmas boxes of nuts contained worms but his Squirrel Brand peanuts did not.

In addition to nut mixes, Squirrel Brand made nougat candies with names like Butter Chews, Nut Chews, Nut Yippees, and perhaps its most famous product, the Squirrel Nut Zipper, a peanut and caramel confection. According to legend, Gerrish named the company after reading a newspaper article about a Vermont man who was inspired to climb a tree after drinking an alcoholic concoction called a "nut zipper."[71] In the 1990s, a swing band named itself after the candy and handed out Squirrel Nut Zippers at its performances. Gerrish's son Hollis, who started working at the company when he was ten and succeeded his father as the firm's head, died in 1997 at age ninety. Squirrel Brand was sold two years later to a Texas company, which moved production out of Cambridge. Although Necco later purchased the rights to the candy, its own bankruptcy in 2018 left its confections in limbo. Clark Bars were bought by the Boyer Candy Company of Altoona, Pennsylvania, while Mary Janes and Squirrel Nut Zippers were not being made at the time of this writing.

The George Close Building on Broadway in Cambridge, 2020. The building served as a manufacturing plant for the George Close Company, makers of penny candies and other novelty sweets, from 1910 to 1939, when it ceased operations. It was converted to apartments in the 1970s. *Source:* Photo by the authors.

Squirrel Brand Salted Nut Co. factory workers. *Source:* Courtesy of the Cambridge Historical Society, Cambridge, MA.

Charleston Chews

The Charleston Chew candy bar, made of vanilla nougat covered in milk chocolate, was created in 1922 by the Fox Cross Candy Company.[72] Its name was meant to capitalize on a highly popular dance called the Charleston, then sweeping the country. By 1931, Fox Cross was located on Cambridge's Landsdowne Street, moving to Candy Row—Main Street—in 1946. The company was sold in 1957 to Nathan Sloane of nearby Belmont, Massachusetts, who ran it until 1980. In the decades the Sloane family owned the brand, they introduced new flavors, doubled production, advertised heavily, and moved the business to a larger facility in the town of Everett, Massachusetts. Though the Charleston Chew is no longer made in Cambridge, it is still an active brand, owned by Tootsie Roll Industries.

Necco Wafers

In 1927, the New England Confectionery Company, better known as Necco, relocated to Cambridge from South Boston and built on Massachusetts Avenue what was called the largest candy factory in the world.[73] There the company produced its signature Necco wafers, thin discs of pastel-colored sugar with flavors such as licorice and clove. The Necco wafer had been invented in 1847 by Bostonians Oliver Chase and Silas Chase, whose patented wafer-slicing machine made the candy possible. During World War II, the US military purchased roughly 40 percent of the company's output to add to soldiers' rations, selecting the Necco wafer because it didn't melt or rot during overseas shipping.[74] In 1901, the company started producing the Sweetheart brand of conversation hearts, pastel-colored heart-shaped candies with romantic messages stamped on them: "Be Mine," "True Love" "Kiss Me." Conversation hearts have been a Valentine's Day tradition ever since. The company manufactured them year round to meet the February demand. After seventy-five years of candy production at the Necco building in Cambridge, the company moved to Revere, Massachusetts, in 2003. Necco was the oldest continuously operated candy manufacturer in the United States at the time of its 2018 bankruptcy and closure.[75]

The New England Confectionery Company (Necco) factory complex at 250 Massachusetts Avenue in Cambridge, ca. 1947. The image is from a twenty-four-count box of Necco's Bolster candy. *Source:* Collection of the authors.

Daggett Chocolates

In 1925, Fred Daggett built a seven-story candy plant at 400 Main Street occupying most of a city block.[76] Daggett produced more than forty brands of chocolates, packaged in boxes and metal tins. The candy bar became the preeminent mode of confectionary sales in the decades after World War II, and many old-line boxed chocolate manufacturers struggled. The company closed around 1960, a few years after Daggett himself died. MIT bought the building almost immediately and has used it for laboratory and office space ever since.

Fig Newtons

In the 1970s, the Nabisco company advertised its fig cookies with a surreal television campaign: An awkward middle-aged man named James Harder, wearing a giant fig costume, performed a brief dance he announced as "the Newton," which he rhymed with "darn tootin."[77] At the close of his performance, Harder stood on one leg ("this is the tricky part," he noted) and belted out "The-Big-Fig-Newton," atonally, at full volume. In 2012, Nabisco's Newtons captured 3.2 percent of the American cookie market, according to the *New York Times*.[78] Though they are no longer baked in Massachusetts, the cookie and its name have local origins.

The Kennedy Biscuit Company began baking in Cambridge in 1839. Frank A. Kennedy expanded the Cambridge plant in 1875, erecting a five-story brick building in Central Square. Kennedy installed reel ovens—somewhat resembling Ferris wheels—with rotating shelves passing through large ovens to permit continuous baking.[79] A Kennedy bakery price list from 1881 included some fifty different types of biscuits, such as soda biscuits, butter biscuits, sugar wafers, and animal crackers.

Kennedy Biscuit merged in 1890 with the New York Biscuit Company and in 1898 with the newly formed National Biscuit Company (Nabisco).[80] In 1892, Kennedy Biscuit first produced its most famous and enduring product. A Philadelphia inventor named James Henry Mitchell developed a machine that could shape a tube of cookie dough and simultaneously fill it with jam.[81] The device resembled a funnel within another funnel, with jam fed into the inner tube and dough fed into the outer one. Continuous ribbons of the filled pastry could be cut into lengths and baked. Mitchell received a patent for his "Duplex Dough-Sheeting Machine" in 1892. Looking for a buyer for his invention, Mitchell persuaded Kennedy Biscuit to try it and sent the machine to the company's Cambridgeport factory.[82] Overseeing the machine's initial installation and operation, Mitchell experimented with different recipes for the jam filling, settling on a fig paste.

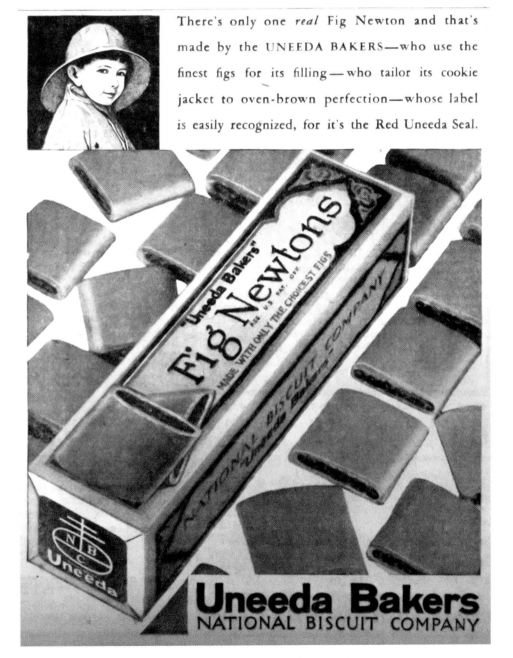

There's only one *real* Fig Newton and that's made by the UNEEDA BAKERS—who use the finest figs for its filling—who tailor its cookie jacket to oven-brown perfection—whose label is easily recognized, for it's the Red Uneeda Seal.

A Fig Newton newspaper advertisement, 1919. *Source:* Collection of the authors.

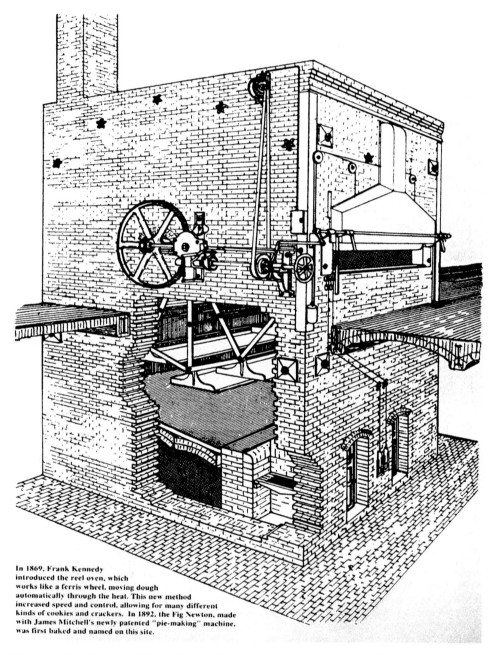

In 1869, Frank Kennedy
introduced the reel oven, which
works like a ferris wheel, moving dough
automatically through the heat. This new method
increased speed and control, allowing for many different
kinds of cookies and crackers. In 1892, the Fig Newton, made
with James Mitchell's newly patented "pie-making" machine,
was first baked and named on this site.

Diagram of a reel oven used in the Kennedy Biscuit Company for continuous baking. *Source:* Photo by the authors.

James Hazen, manager of the Kennedy Biscuit bakery in the 1890s, is credited with naming the fig-filled cookie for the nearby town of Newton, Massachusetts.[83] Eventually, the word *Fig* was added to the cookie's name, which became Fig Newtons (though *Fig* was dropped from the name in the 2010s as part of a rebranding). Newtons were originally sold in bulk; eventually Nabisco began packaging them in colorful paperboard boxes in convenient quantities for household consumption.

The Kennedy Biscuit Company bakery at 129 Franklin Street in Cambridge remained in operation through the 1940s, when Nabisco moved production to a newly built plant in New Jersey.[84] The bakery building was used as a shoe factory for another three decades and in 1989 was converted to an apartment complex called the Kennedy Biscuit Lofts.

For over 150 years, Cambridge thrived as a center for candy and cookie making. Companies here developed new brands and responded to changing tastes with

The former Kennedy Biscuit Company factory near Central Square, where the Fig Newton was first baked, 2015. The building was converted to loft apartments in 1989. *Source:* Photo by the authors.

technological innovation and clever marketing. A critical mass of candy manufacturers developed locally, supported by infrastructure such as the nearby Revere Sugar Refinery Company and available plots of industrial land. The factories were staffed with recent immigrants and their children. Skilled workers could begin their careers at one company and end at another, sometimes as the owner.

After World War II, this critical mass shrank, and the local candy industry lost its competitive advantage. Some of the companies that survived moved to the Midwest to be closer to raw materials like corn syrup or to the suburbs for cheaper labor and land. The almost total loss of its candy-making industry changed the fabric of Cambridge—even the way it smelled.

But it also proved essential to the city's emerging role as a biotechnology hub. In 1974, MIT biologists founded the Center for Cancer Research in a renovated chocolate factory on the corner of Main and Ames streets. Researchers there made profound insights into the microbiology of cancer, beginning to unravel for the first time what happens when a tumor invades the body. At a certain point, the scientists began crediting the building for their successes. Their cramped quarters were often the source of complaints until the center relocated four decades later, but the closeness also meant that people with different specialties literally bumped into each other regularly and ended up sharing ideas. The concepts of cross-disciplinary research and the importance of casual interactions are now foundational in science, with new labs deliberately designed around these ideas.

In 2002, Swiss pharmaceutical giant Novartis chose to locate its global research and development headquarters in Cambridge, in part because corporate executives believed that such casual interactions—call it the *coffee shop effect*—would improve their science. The building that they picked to form the centerpiece of their new R&D headquarters and that sparked a renaissance in Cambridge was the former Necco candy factory.

POLAROID AND THE ORIGINS OF INSTANT PHOTOGRAPHY

For more than three decades, from the late 1940s to the 1970s, the Cambridge-based Polaroid Corporation was recognized as one of the most innovative consumer products companies in the United States. Polaroid's instant cameras and film systems were some of the most sought-after products in postwar America, beginning with an instant monochrome camera in 1948 and later including the distinctive white-bordered color photos that are universally known as "Polaroids." At its peak in the 1970s, Polaroid reported more than a billion dollars in annual revenues, employed tens of thousands of people, and had no real competitors.[85] The company had a broad impact on American culture, arts, and business. Polaroid's success inspired Steve Jobs when he was building Apple, which later took on a similar role as a consumer products–focused innovator.[86]

Polaroid was the singular vision of a driven man, Edwin Land, who dropped out of Harvard—twice—and then built an empire a few miles across town. For much of his time at its helm, Land managed Polaroid from an office and laboratory at the intersection of Main and Osborn streets in Cambridge. From his office, "he would open one door and there was this laboratory, full of test tubes and Bunsen burners and whatever he needed," says Robert Alter, a Polaroid employee from the early 1980s to the mid-1990s.[87]

Land was part eccentric inventor and part showman. In the weeks before the company's annual meeting with investors, where new products were often unveiled, thousands of employees across the company were mobilized "to do whatever Land said he wanted done," says Alter. Crews worked through the night to build sets and prepare special effects. "It was a big point of pride that he could do that—that all of these folks were available to him. . . . He could take his science and dazzle Wall Street with it. He could hold this giant room spellbound while he outlined the plan for his upcoming product and have everybody on the edges of their seats anticipating what miracle he was going to pull out of his hat."

Beginnings

Edwin Land was born in Bridgeport, Connecticut, in 1909 and entered Harvard at the age of seventeen. As a student, he decided to study the properties of polarized light—the tendency of light waves to vibrate along an axis perpendicular to their path of travel.[88] Land stunned his parents when he told them that he was not planning to return for his second year at Harvard, instead moving to New York City to continue his experiments on polarized light. Working in a shabby basement room and—surreptitiously—a Columbia University laboratory, Land developed a plastic sheet embedded with iodine crystals that proved capable of polarization. He returned to Harvard in 1929, where he met physics instructor George Wheelwright.

In June 1932, Land again left Harvard before completing the requirements for a degree. With Wheelwright, he set up a laboratory on Mount Auburn Street in Cambridge, intending to develop his polarizing material into marketable products.[89] The business grew slowly at first. In 1934, the Eastman Kodak Company placed a $10,000 order with Land-Wheelwright for polarizing camera filters, launching the company toward financial stability.[90] Soon they won another large contract to produce polarizing material for the American Optical Company, a maker of sunglasses.[91] In 1937, Land-Wheelwright Laboratories was renamed Polaroid Corporation after its signature product, following an infusion of capital from Wall Street. In 1940, Polaroid settled in a five-story factory building at 730 Main Street in Cambridge. The move brought Polaroid close to the MIT campus, from which it would draw many future employees, but the new headquarters was more directly surrounded by factories making soap, rubber, and candy.

By the end of 1940, Polaroid faced its first financial crisis, the result of a declining market for its sunglasses and its failure to sell car makers on the idea of polarized headlights. As the US military ramped up preparedness for World War II, Polaroid was saved by a US Navy contract for an optical device called a Position Angle Finder.[92] Soon the military was buying the company's polarizing goggles for pilots. Land was convinced that, for the Allies to win the war, American scientific and technical talent needed to be fully mobilized, and he threw his company fully into government service. Over the course of World War II, Polaroid applied its research skills to the development of optical rangefinders, three-dimensional aerial photography, heat-seeking weapons, a machine-gun trainer, and synthetic quinine. In 1941, the company reached sales of $1 million; several years later it received a navy contract for $7 million.

As the war dragged on, Land began preparing Polaroid for peacetime. The company had top-notch researchers and mechanics, but he questioned how best to apply their talents to new, commercial ends. As the story has been often told, Land was inspired during a December 1943 trip to Santa Fe, New Mexico, with his family, a rare break from his work. His three-year-old daughter Jennifer accompanied Land as he took photographs of the Santa Fe scenery with his Rolleiflex camera. "Why can't I see them now?" she asked her father.[93] Land had no immediate answer, but he spent the remainder of the day mapping out in his head the technological hurdles that would need to be overcome to create an instant photograph. He later recalled:

> As I walked around that charming town I undertook the task of solving the puzzle [Jennifer] had set me. Within the hour, the camera, the film, and the physical chemistry became so clear to me that with a great sense of excitement I hurried over to the place where [patent attorney] Donald Brown was staying, to describe to him in great detail a dry camera which would give a picture immediately after exposure.[94]

His concept of instant photography would take another five years to bring to market. The first Polaroid Land Camera made its debut the day after Thanksgiving 1948 at the Jordan Marsh department store in Boston. The camera, called the Model 95, sold for $89.75 (the equivalent of about $1,000 in today's dollars); an eight-exposure package of film cost $1.75.[95] It produced completely dry, sepia-toned monochrome images in sixty seconds. On the day of the camera's introduction, Jordan Marsh sold all fifty-six of the Model 95s it received from Polaroid—which were all the cameras that Polaroid had

The Polaroid Model 95A Land Camera, introduced in 1954 as an updated version of the original instant photography system launched in 1948. *Source:* Collection of the authors, photo by the authors.

gotten from the subcontractor making them. The Polaroid camera became a must-have item for increasingly affluent Americans in the booming postwar years. Instant photography drove explosive growth at Polaroid. In 1960, the company posted sales of nearly $100 million dollars and employed nearly three thousand workers. A decade later, Polaroid reported $508 million in sales and more than ten thousand employees.[96]

In the 1950s and 1960s, Polaroid expanded its footprint in Cambridge, occupying many former factory buildings near the MIT campus.[97] In an era when the city of Cambridge was staggering from a rapid loss of manufacturing jobs, Polaroid represented a bright spot, a manufacturing company that was actually growing and hiring.

Working at Polaroid

At times, Land worked himself and his closest assistants to the point of exhaustion. He demanded unflinching personal loyalty from his lieutenants. Land carefully cultivated an image of himself working for days, sometimes even weeks at a time, with people bringing him sandwiches or ordering takeout food "to keep him going," Alter says. Land would call employees in the middle of the night, saying he needed them urgently. "That legend was so powerful at Polaroid: This brilliant driven genius, who could just go for weeks at a time hardly eating or sleeping. He probably did that for a long time, and he loved that image of himself, too," Alter says. "He constructed the company in that culture. It must've been tremendously exciting."

Despite Land's personal work ethic, the company became known for its progressive employment policies. Land articulated dual aims for the company (as recalled by a vice president of the company, Marian Stanley, in a 1996 oral history): "One, to create useful and exciting products, and secondly, to bring the best out in people and give them useful, exciting work."[98] Polaroid gained a reputation for hiring young women in research roles at a time (the 1940s) when it was not commonplace elsewhere. The company offered generous maternity leave, eliminated time clocks for its workforce, and encouraged employees to pursue continuing education, with courses offered within the company on topics such as injection molding, basic math, and chemistry.[99]

Land sought out talented people regardless of their backgrounds and often gave hires considerable leeway in their job responsibilities. "You didn't have to be credentialed," says John Reuter, who had just completed a master's degree in fine arts when he was hired into a technical research lab at Polaroid.[100] "They didn't care. They wanted strong individuals that had an open mind and a creative mind, and they always felt they could train you in whatever path they wanted you to go into."

One example was Meroe Morse, described as "a cheerful, intelligent, energetic young woman who had come to his laboratory in 1944, straight from Smith [College].

Portrait of Edwin Land. *Source:* Courtesy, MIT Museum.

700 Main Street in Cambridge, which once held the office and laboratory of Polaroid founder Edwin Land, 2020. *Source:* Photo by the authors.

There she majored in art history," according to Land biographer Victor McElheny.[101] "Although she took not a single course in physics, chemistry—or business administration—she, like others from Smith, showed aptitude in Land's laboratory, stimulated by Land's insistent mind." From the 1940s to the 1960s, Morse oversaw the development of Polaroid's black-and-white photography, turning many of Land's "impossible" ideas into tangible products.

Former Polaroid executive Marian Stanley recalled another example, where Land put his trust in an employee: "Howard Rogers was an automobile mechanic that Dr. Land thought was particularly skilled," she said. "He gave Howard a lab and money and told him to go away and develop color photography. . . . Rogers went away. . . . He came back some years later with not much heard from in-between times, and said 'I think I've done it.' And he had done it."[102]

SX-70: Realization of a Thirty-Year Dream

In the early 1960s, Land launched a program to develop an instant color camera called the SX-70. The goal was to build a camera that would, using advanced chemistry and

Polaroid machine shop workers, ca. 1938. *Source:* Polaroid Corporation records, Baker Library, Harvard Business School.

electronic exposure controls, create and then eject a color photographic print, with nothing to peel off or throw away. Using the considerable profits from its older film systems, Polaroid invested more than $500 million in developing the fundamental chemistry, camera systems, and even factories needed to make Land's dream a commercial reality.[103] The project started with the development of a new type of "integral" film, containing thirteen layers, each serving a separate function.

With the introduction of the SX-70 camera and film system in 1972, Polaroid shifted from being primarily a research laboratory to a fully integrated manufacturing business. Rather than contracting other firms to make its camera and film components and chemicals, Polaroid built or expanded factories of its own to complement the research headquarters in Cambridge. After difficulties with a battery supplier, Polaroid began to manufacture the batteries built into its film packs to drive camera motors and electronics. By 1978, largely on the strength of the SX-70 camera and film, Polaroid had annual sales of nearly $1.5 billion.

Polavision and Land's Departure

After the introduction of the SX-70, Land became obsessed with creating a system for instant home movies. For years, photographers talked about Polaroid's movie project and the ways it would revolutionize their industry, Robert Alter says. Repeated delays just boosted the enthusiasm about how good it would eventually be. Alter recalls being a young graduate student at MIT in 1978 when the long-anticipated Polavision finally arrived. "I remember one day walking into a camera store, and there it was. It had this great display—this highly designed, expensive kiosk where you could test out and see the movies," he says, recalling his excitement. "This is great. This is going to change everything." He asked for a demonstration. The picture was hard to see, grainy and of poor quality. "I thought, 'Oh, my God, this is a disaster, this is terrible.'" A total disappointment, he says. "Everybody's been expecting this amazing thing, and it just wasn't very good."[104]

Polavision sold poorly. The company discontinued it in 1979, chalking up tens of millions of dollars in losses, writing off unsold inventory and manufacturing equipment, and eliminating more than 2,500 jobs.[105] For Land, Polavision represented a huge gamble on the company's future and a significant failure. He left as chief executive in 1980.

Land's successors in the 1980s and 1990s faced challenges that slowly sapped the company of the revenue and profits it would need to innovate its way out of difficult times. Alter says he tried repeatedly to talk to executives about getting into digital photography. But the company that had relied on adaptability, flexibility, and innovation was now living off of its aging patents. It couldn't figure out how to gain a new competitive advantage. "Polaroid's incredible success was also its downfall," Alter says. "Land had built this amazing patent empire that created a product that people wanted and was incredibly profitable, but then eventually the patents [ran] out and technology changed. The secret to their success was also what inhibited them and kept them from being able to adapt and change."

The introduction of one-hour film processing labs for noninstant film, the increasing popularity of 35 millimeter cameras, and the advent of digital photography ate away at Polaroid's sales. The company was unable to generate new must-have products; its vaunted reputation for innovation was seriously damaged.

Mounting debt and a changing marketplace added to its problems. Polaroid filed for bankruptcy in October 2001 and soon left its Cambridge headquarters. The reorganized company stopped producing instant film in 2008, leaving it with no more than 150 employees and depriving Polaroid camera owners of film. In 2009, a group of investors, former employees, and Polaroid enthusiasts calling themselves The Impossible Project purchased Polaroid's factory in the Netherlands to continue film production and introduce new instant cameras. After a series of restructurings and ownership changes, the Polaroid brand lives on but without a Cambridge connection.

THE HI-LITER MARKER

By the end of the 1950s, manufacturers had begun to leave Cambridge. The companies that remained, like Carter's Ink, had to innovate to survive. Started in Boston in 1858, Carter's moved to Cambridge in 1910, where it built a factory on First Street facing the Charles River.[106] The company had long since learned the importance of product innovation. In the 1860s, Carter's thrived because it developed a competitive advantage: its Combined Writing and Copying Ink could be used both for normal writing and, when paired with a special paper, to make multiple copies of a document. This was long before the invention of the photocopy machine or even carbon copies. John W. Carter resuscitated the company after an 1872 fire left it with nothing but its reputation, formulas, and a successful salesman. Carter hired a skilled chemist to develop new ink formulas. By 1884, the company was manufacturing nearly five million bottles of ink a year.[107] Carter's continued to innovate, adding photo library paste, ink eradicator, new ink colors (including gold, silver, and white), and, as typewriters became popular, carbon paper and typewriter ribbons.

In its Cambridge plant, Carter's installed the latest automated equipment for filling, capping, and labeling bottles. In 1926, the company's First Street building was topped with an electric clock, becoming a landmark on the riverfront.[108] By 1930, Carter's was one of the largest ink producers in the United States.

The company remained profitable through World War II and continued to expand through the 1950s. Its motto was "there is nothing so good that it can't be better."[109] Ink bottle collectors Ed and Lucy Faulkner wrote of Carter's in 2003: "Every time things got rough they came up with new products to keep the business going."[110]

In March 1958, the company introduced a line of felt-tipped permanent markers called "Marks-A-Lot" that proved instantly successful.[111] Carter's sold Marks-A-Lot in grocery and variety stores, producing them in bright colors and displaying them in easy-to-see clear plastic blister packs.[112] They were particularly a hit with children.

The Carter's Ink Factory in Cambridge, 1968. Industrial chemist Francis Honn worked in this complex when he invented the yellow highlighting marker in 1963. *Source:* Carter's Ink Collection, Cambridge Historical Commission.

Parents enjoyed the creativity but not the fact that Marks-A-Lot didn't wash off walls, clothes, or furniture. Mothers wrote to the company begging for washable versions. In response, Carter's chemists formulated inks made with water rather than oil or solvents and sold them as "Draws-A-Lot." The company expanded its manufacturing plant to meet demand for the new markers.

In 1959, Carter's hired Francis Honn, an industrial chemist then working for 3M in Minnesota. Honn was responsible for quality control, and he often grabbed a few samples from the assembly line. One day, he tested a translucent yellow marker by drawing it across a page of type. "The black type literally jumped out at me," Honn recalls in an unpublished memoir nearly a half century later.[113] "Here was a better way of highlighting key printed words than underlining!"

Honn instructed the company's sales staff to send out letters to customers who had complained or asked questions, writing over a few key words, like "refund" and "free sample," with yellow ink. At the bottom of the page, Honn added a sentence: "Wonder

Francis Honn, who invented the Hi-Liter marker while employed at Carter's Ink in Cambridge. *Source:* Courtesy of the Honn family.

where the yellow came from? Carter's Reading Hi-Liter"—riffing on a popular toothpaste advertisement that claimed "You'll wonder where the yellow went when you brush your teeth with Pepsodent."[114] Within a week, Honn said, "letters came pouring in asking where they might buy this wonderful product."[115] Honn received permission from the company's executive board to sell the Reading Hi-Liter in a test market; company president Nate Hubley suggested school bookstores, which already carried Carter's products. The marketing department designed an eye-catching blister pack, pairing a Draws-A-Lot pen with a Hi-Liter.

As Honn remembers it, the company had to invest only $3.50 to make the new markers, creating a new rubber plate to print "Reading Hi-Liter" on their existing yellow markers. Honn and his team didn't so much "invent" the highlighter as recognize its usefulness and find ways to sell it to the world.

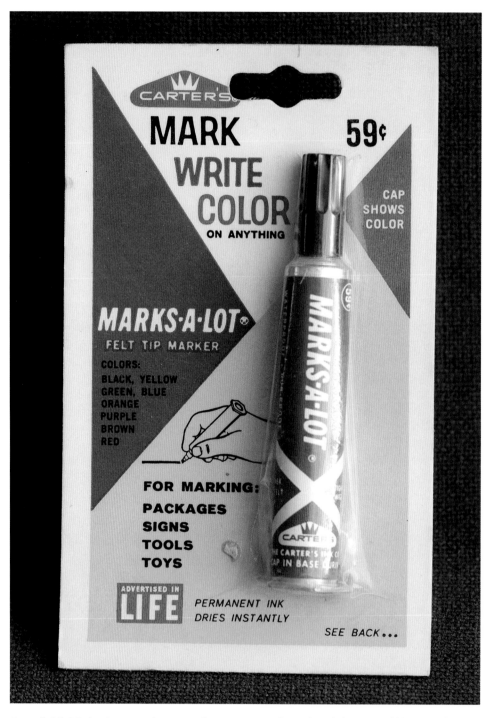

Carter's Ink Marks-A-Lot marker, ca. early 1960s. Carter's employed this type of blister packaging to sell its first highlighting markers. *Source:* Collection of the authors.

Highlighting markers have changed the way that readers take notes, editors revise texts, and students study for exams. Highlighting markers now come in pink, green, orange, and blue in addition to the original yellow. ("Yellow is still the only good color," Honn argued in 2009, basing his assessment both on his own perceptions and on the theory, which he'd learned at 3M, that yellow and black offer more contrast to the eye than other color combinations.) According to a *New York Times Magazine* story in 2012, 85 percent of markers sold are either yellow or pink.[116] Even on a computer, an icon of a yellow felt-tipped marker is used as a symbol for highlighting text in programs like Google Docs, Adobe Acrobat, and Microsoft Word.

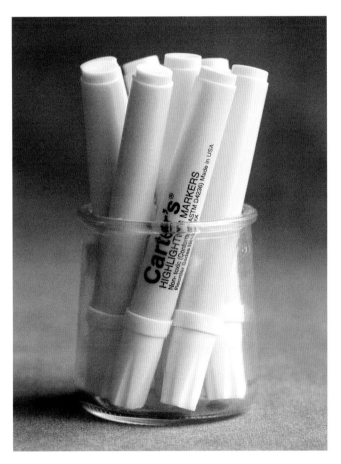

Carter's Highlighting Markers were still marketed by the Avery Dennison Company in the early twenty-first century, though in a different form than in the 1960s. *Source:* Photo by the authors.

Honn left Carter's Ink by 1970. After a long career as a chemist and company executive, in which he was granted more than twenty-five patents, Honn retired in 1988 and died in New Jersey in 2016 at age ninety-four.[117]

Unfortunately, the Hi-Liter marker was not enough to ensure Carter's independent future or its continued presence in Cambridge. In the 1960s, the company opened large manufacturing plants in Tennessee and North Carolina.[118] After its purchase in 1976 by the Dennison Manufacturing Company, an office supply and label maker, Carter's Ink moved out of Cambridge completely. For many years, the Carter's name remained on packages of ink, Hi-Liters, and stamp pads made by the Dennison (now Avery Dennison) company.

As a sign of its ubiquity, the highlighting marker even has its own Wikipedia entry.[119] The word *highlight* has become a verb. "The fact that something so simple could still have such an impact today makes me happy," Honn said in a 2016 interview.[120]

During Harvard College's first century, from 1636 to about 1750, the curriculum was focused on ancient languages (Latin, Greek, and Hebrew), logic, public speaking, and arithmetic. In its early decades, roughly half of Harvard graduates became Puritan ministers. There was little emphasis on the creation of new knowledge. But by the dawn of the American Revolution, natural philosophy—the origins of what today we call science—had grown as an area of study. In 1766, Harvard began to assemble and exhibit a collection of scientific artifacts and art works called the Philosophy Chamber.[1] The study of fields like astronomy, algebra, and physics soon followed. Harvard's Medical School was founded in 1782.

In 1799, **Benjamin Waterhouse**, a Cambridge physician and one of the first members of Harvard's medical faculty, successfully vaccinated members of his household against smallpox, a highly contagious and often deadly disease. Waterhouse learned of the vaccination technique from its inventor, Englishman Edward Jenner, and his family members were the first Americans to benefit from it. Waterhouse later helped spread the practice of smallpox vaccination across the country.

Over the next half century, scientific advances became a means for the young nation to declare its intellectual independence from Europe. "The center of gravity of science starts to shift from the old [world] to the new," notes Jonathan McDowell, an astronomer at the Harvard-Smithsonian Center for Astrophysics. Public excitement for **astronomy** in America grew following the appearance of the Comet of 1843. Building on that enthusiasm, in 1847 Harvard was able to establish a new astronomical observatory on Garden Street in Cambridge. Its German-made telescope, called the Great Refractor, was used to capture some of the first clear photographs of the moon and stars.

The 1840s were an auspicious decade for the advancement of the natural sciences in Cambridge. **Asa Gray**, the preeminent American botanist of his era, joined the Harvard faculty in 1842. Gray's correspondence with Englishman Charles Darwin about the geographic spread of botanical species helped cement Darwin's nascent theory of natural selection.

A Prospect of the Colledges in Cambridge in New England by Sidney Lawton Smith (1845–1929), ca. 1917. The engraving is based on the 1726 William Burgis drawing of the Harvard College campus. *Source:* Prints and Photographs Division, Library of Congress, Washington, DC.

In 1847, Harvard established the Lawrence Scientific School, endowed by textile magnate Abbott Lawrence to advance the practical uses of science. Two of the school's first hires were **Eben Horsford**, a specialist in agricultural chemistry who greatly improved baking powder, and **Louis Agassiz**, a popularizer of science in America known for his contributions to the fields of zoology and geology but also for his controversial and unscientific views on race.

Local support for the sciences got another boost in 1869, when Charles William Eliot, a chemist, took on Harvard's presidency with an agenda to reshape the university in the model of the great German research institutions. During his forty-year tenure, Eliot transformed Harvard from a provincial liberal arts college to a preeminent research university.

In 1895, Eliot asked a young physics professor, **Wallace Sabine**, to investigate the acoustical failings of a newly built lecture hall at the university's Fogg Museum. In the course of answering Eliot's question (a process that took him several years of painstaking research), Sabine largely invented the field of architectural acoustics, which allows building designers to understand how sound travels inside a space and to manipulate it to their advantage.

In 1916, the Massachusetts Institute of Technology moved from Boston's Back Bay to Cambridge's Charles River waterfront, on land created by dredging the river and dumping the spoils behind a stone embankment. With funds donated anonymously by photography pioneer George Eastman, the founder of Eastman Kodak, MIT built a campus of impressive neo-Classical buildings. Over the course of the twentieth century, MIT grew from a vocationally oriented institute to an internationally recognized leader in science and technology research. During World War II, MIT was the single largest recipient of federal government research grants. The links forged during the war years between government funding agencies and Cambridge's academic institutions have supported the growth of scientific research in Cambridge ever since.

In 1914, T. W. Richards won Harvard's first Nobel Prize for work he began as a graduate student, fixing the atomic weights of chemical elements. Harvard faculty members have since been awarded dozens of Nobel prizes for groundbreaking research in topics from global poverty alleviation to the discovery of smell receptors.[2] MIT professors have won six Nobel prizes in the field of economics alone since Paul Samuelson took home a medal in 1970,[3] and the school claims sixty-five Nobel recipients of all types won by faculty, staff, and students.[4]

Though Cambridge's reputation as an academic research powerhouse has been several centuries in the making, few cities in the world today can compete with its record of scientific achievement.

BENJAMIN WATERHOUSE: VACCINE PIONEER

On July 8, 1799, Cambridge physician Benjamin Waterhouse made a small incision in the arm of his five-year-old son Daniel.[5] He inserted a bit of thread that had been infected with cowpox, a virus that caused relatively mild disease in humans, and covered the wound with sticking plaster—the eighteenth-century equivalent of a Band-Aid. He then performed the same procedure on a twelve-year-old servant boy. Some days later, when Daniel's arm was covered with the oozing sores characteristic of cowpox, Waterhouse pressed a particularly liquid sore onto a cut he'd made in the arm of his three-year-old son Benjamin. Then, when sores began to appear on the arm of young Benjamin, Waterhouse used them to infect the boy's one-year-old sister, her nursery maid, and his seven-year-old sister, Elizabeth. For each, Waterhouse recorded the swelling, pain, and sores that occurred, noting that they were all precisely what he had been told to expect, tolerable in all of his patients excepting the servant boy, who "treated himself rather harshly by exercising unnecessarily in the garden, when the weather was extremely hot (Farht.Thermr. [sic] 96, in the shade!)."

Waterhouse was investigating whether exposure to cowpox would provide the members of his household with immunity to smallpox, a much more virulent disease that killed or blinded many of its victims and left survivors covered with scars. To test that hypothesis, Dr. William Aspinwall, who ran a smallpox inoculation hospital in nearby Brookline, brought Waterhouse's servant boy to his hospital and placed a bit of smallpox taken from an active patient into each of the boy's arms.[6] Four days later, the boy's arms were clearly infected and sore. But the disease never progressed, and in a day or two, his arms began to heal. On the twelfth day after the experiment, he was allowed to go home, the effectiveness of his vaccination proved. "One fact, in such cases, is worth a thousand arguments," Waterhouse concluded.

The boy, who would remain unidentified, was the first person in America confirmed to have been protected from smallpox through a vaccine. Waterhouse, one of the founding professors of Harvard Medical School, brought smallpox vaccination to the United States and later helped spread it to the nation.

Born in 1754 in Newport, Rhode Island, Waterhouse began training by his midteens for a career in medicine.[7] In 1775, he left for Europe to further his medical education—on what was reportedly the last ship to sail from Boston Harbor before the start of the American Revolution.[8] He remained abroad for the duration of the war, although, raised as a pacifist Quaker, he would not have participated in the fighting had he stayed.[9] His absence proved both helpful and harmful to his later career.

Despite spending the war abroad, Waterhouse was clearly an American patriot. In 1778, signing the matriculation records at the University of Leyden in the Netherlands, where he would study for more than four years, Waterhouse wrote in Latin that he was a "Citizen of the Free and Independent State of America."[10] After graduation, he visited Paris and was invited to dinner by Benjamin Franklin, then the American minister to France and a close friend of his great uncle's.[11] John Adams, who had been sent to Europe by the Continental Congress to negotiate peace with England, was also at the table. Waterhouse cemented his friendship with Adams once he returned to Holland, where he studied for an extra year.

On his return to the United States in the early 1780s, Waterhouse was America's best-educated physician. Harvard University hired him in 1782 as one of the first three

B. Waterhouse, M.D., Professor of the Theory and Practice of Medicine, engraving by S. Harris.
Source: Courtesy, Center for the History of Medicine, Harvard Medical School.

professors at its new medical school.[12] Waterhouse was also appointed to a lectureship on natural history, which then included biology, mineralogy, and geology.[13]

Waterhouse's years back in America weren't easy. He was treated with hostility by members of Boston's medical and scientific community because of his absence during the war and because he was an outsider both by his Newport birth and his Quaker ideals. Also, as one biographer put it, "he assumed an attitude of insufferable superiority."[14] He was unable to establish a successful medical practice in Boston and was forced to relocate to Cambridge as early as 1787, according to biographer Philip Cash.[15] Harvard came to Waterhouse's financial rescue on several occasions and finally, in 1793, purchased and rented him a house in Cambridge at the north end of the Common on a road later renamed and still called Waterhouse Street.[16] Though it had been founded more than a century and a half earlier, Cambridge remained rural, housing just 2,115 inhabitants, mostly south of the Common.[17] Moving to Cambridge alienated Waterhouse further from Boston's medical establishment, and his influence declined still more, Cash wrote.

Even as a professor, Waterhouse did not always get the respect he felt he deserved. Oliver Wendell Holmes, the famous Harvard-educated doctor and poet, once said of Waterhouse (who vaccinated him):[18] "The good people of Cambridge listened to his learned talk when they were well, and sent for one of the other two doctors when they were sick."[19]

In early 1799, Waterhouse received a book as a present from a London friend.[20] Its author, Englishman Edward Jenner, had recognized that milk maids who caught a mild disease called cowpox received the same immunity to smallpox as if they'd had the devastating disease themselves. By deliberately infecting people with cowpox, he could protect them against smallpox.

At the time, smallpox was a terrifying, highly contagious scourge that killed one in six patients.[21] In a 1721 outbreak in Boston, smallpox killed more than 850 people and sickened six thousand—more than half the city's population.[22] Nearly eighty years later, the outbreak still loomed large. The only known protection was inoculation: being deliberately infected with the disease in hopes of contracting a mild case. One in 350 people inoculated by doctors like Aspinwall died from the infection, and they were all contagious while sick.[23]

Waterhouse quickly saw the potential for Jenner's vaccine—which was safer than inoculation and not contagious. In 1799, he presented Jenner's research to the American Academy of Arts and Sciences, which met at Harvard and was presided over by his friend John Adams, then the US president.[24]

By early July, Waterhouse had received a supply of the vaccine from England and began using it, first on his family and then, when he was convinced of its effectiveness, on others. The vaccine was delivered differently than those we receive today. Instead of a carefully calibrated dose injected with a syringe, a doctor would do as Waterhouse had, making a cut in a patient's arm and inserting a thread that had been coated with cowpox. A few days later, pustules would appear near the wound site. Timed correctly, a doctor could take material from these pustules to infect other people. Even Waterhouse struggled to master this process. Many of the people he and others vaccinated that first year did not get protection from smallpox.[25]

Initially, Waterhouse tried to exploit his status as the only person in the United States with the vaccine material by tightly controlling how it was circulated. "He did not incur others' enmity because he inaugurated the practice, but because of the way he did it," wrote one biographer.[26] Though he was accused of trying to profit from the misery of others, he said he was worried about the potential for disaster.

The house of Benjamin Waterhouse on Waterhouse Street in Cambridge. *Source:* Courtesy, Cambridge Historical Society.

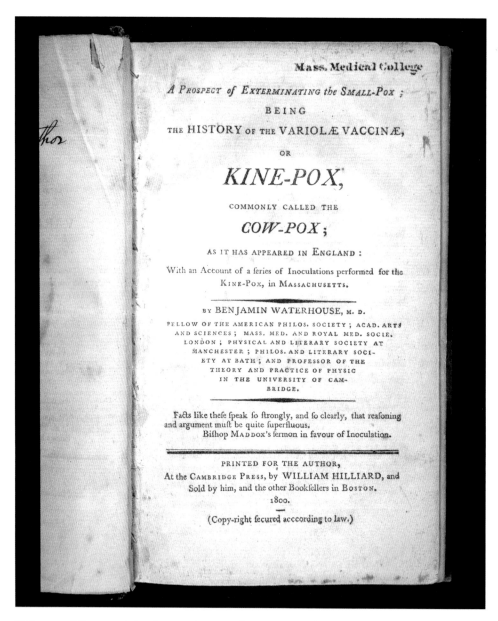

Title page of Benjamin Waterhouse, *A Prospect of Exterminating the Small-Pox; Being the History of the Variolae Vaccinae or Kine-Pox, Commonly Called the Cow-Pox; as It Has Appeared in England: With an Account of a Series of Inoculations Performed for the Kine-Pox, in Massachusetts* (Cambridge: Cambridge Press, 1800). *Source:* Courtesy, Center for the History of Medicine, Harvard Medical School.

Letter from Benjamin Waterhouse to Thomas Jefferson, November 5, 1801. *Source:* Courtesy, Center for the History of Medicine, Harvard Medical School.

The Waterhouse family tea set featured images of cows (a reference to cowpox, the less virulent virus used to formulate the smallpox vaccine). *Source:* Courtesy, Center for the History of Medicine, Harvard Medical School.

Sadly, he was proven correct. In early October, Dr. Elisha Story, a practitioner in nearby Marblehead, vaccinated his daughter with cowpox. Two weeks later when she developed symptoms, Story used her pustules to pass the disease onto others.[27] Unfortunately, she had instead developed smallpox, and the virus took hold in his seaside town, eventually killing sixty-eight people. Waterhouse changed his approach after that first year and never again charged money for the vaccine or his knowledge.

Eager to spread word of his success, Waterhouse wrote to President Thomas Jefferson, who said by return mail that he already knew of Waterhouse's work and offered effusive praise of his efforts. "Every friend of humanity must look with pleasure on this discovery, by which one evil more is withdrawn from the condition of man," Jefferson wrote on December 25, 1800. Six months later, Waterhouse sent Jefferson some infected thread with detailed notes on its proper use. With Jefferson's support, the vaccine was soon distributed in Washington, DC, Maryland, Virginia, and other southern states. Doctors in New York and Philadelphia, who worked independently from Waterhouse, had struggled with the vaccine, and it wasn't until they used his instructions that they were able to successfully vaccinate people in those cities.

Despite his national success, Waterhouse remained an outcast in Boston. After years of bitter personal and political disputes, he was pushed out of Harvard in 1812 for "embarrassing the affairs of the medical institution" and lying to the corporation.[28] First Jefferson and then President James Madison came to his rescue by offering him well-paid medical posts with the US military in New England.[29] Waterhouse died in 1846 at age ninety-two. The last natural outbreak of smallpox in the United States was just over a century later, in 1949,[30] and in 1980, the World Health Organization declared that vaccination had eradicated smallpox worldwide.[31]

MAPPING THE HEAVENS: ASTRONOMY IN CAMBRIDGE

Walking up the curved stairwell toward the Harvard College Observatory's Great Refractor telescope is like stepping back in time. A marble bust sits in a niche along the stair. A creaky door opens at the top with a large key, and a steep threshold places the visitor into a huge, domed room. The wood-and-brass telescope, dating to 1847, nearly touches a slice of roof that slides open, allowing the device to peer into the night sky. Behind the telescope, a large metal frame cradles an upholstered double seat, looking like something from a steampunk novel or a vintage movie theater. From a sitting position, an astronomer can crank the seat up or down to get level with the eyepiece. More than 170 years ago, men occupying that seat captured some of the first clear images of the heavens. Later, women working in rooms downstairs used photographic images made with the telescope to catalog all of the stars in the sky, looking for changes that revealed movement, stellar relationships, and long-ago explosions. The stretch of Garden Street in Cambridge known as Observatory Hill became an important center of astronomical research in the nineteenth century, a role it continues to play today.

The Harvard Observatory and the Beginnings of Astrophotography

Mariners had long studied the planets and stars, and poets had long written about them, but detailed understanding of the heavens prior to the early 1600s was stymied by the limits of human vision. Then optical telescopes were invented in Europe. As early as 1642, students at Harvard College could study astronomy, though the college did not own a telescope until perhaps the 1670s.[32] In 1839, Harvard College established an observatory at the foot of what is now Quincy Street and appointed William Cranch Bond, a Boston clockmaker and amateur astronomer, as its head. At his own home in Dorchester, Bond had cut a hole in the ceiling of his parlor to fit his telescope. Moving to Cambridge, Bond was required to use his own equipment, and he served at first in an unpaid position.

The Harvard College Observatory in Cambridge, ca. 1900, Detroit Publishing Co.
Source: Courtesy of the Library of Congress, Washington, DC.

By the 1840s, pressure was building for Harvard to place more emphasis on astronomical research, says Jonathan McDowell, an astrophysicist at the Harvard-Smithsonian Center for Astrophysics.[33] The university operated "a bunch of small telescopes on the roof of Massachusetts and Dana halls," McDowell says, but "Cincinnati was building a big telescope, and people in New England went: 'Wait a minute!'"

The passage of the Great Comet of 1843 created widespread public enthusiasm for astronomy. In the Boston area, the comet "got them all stirred up through panic," says Owen Jay Gingerich, emeritus professor of astronomy at Harvard.[34] Bostonians were afraid not of the comet, Gingerich says, but of being seen as an intellectual backwater because the region lacked a large telescope.

Former President John Quincy Adams was among those who began lobbying Harvard alumni—"all these great names of New England"—to raise funds to buy a larger telescope, McDowell says. Donors included founders of New England's textile industry and the ice baron Frederic Tudor. In all, they raised more than $25,000 for the purchase and installation of the 15-inch Great Refractor telescope. Its manufacture in Munich, Germany, would take several years.

This was a key time for science in the young nation. "It's the moment when the US stops thinking of itself as a colony and starts to think of itself as a superpower—and so is starting to invest in big science," McDowell says, adding that Harvard's telescope might have been the first major American investment in big science. "If that's true, that's the moment when the center of gravity of science starts to shift from the old [world] to the new. It's much bigger than Harvard," he says.

In 1844, flush with public support and awaiting the new telescope, Bond moved his observatory to the highest point in Cambridge, what became known as Observatory Hill. His new neighbor there was the botanist Asa Gray, who gave Bond plants to grow on the Observatory grounds. When the Great Refractor was installed in 1847, it was the largest telescope in the United States, and it remained so for twenty years.[35] Bond's son George Phillips Bond was appointed "assistant observer" in 1846. (He became director of the Observatory on his father's death in 1859).

In 1849, the Bonds and John Adams Whipple made a remarkably clear and detailed photograph of the moon—the earliest known surviving moon photo ever taken, McDowell says. Two years later, the image was exhibited at London's Crystal Palace Exhibition to much acclaim.

Meanwhile, George Bond and Whipple had captured the first, albeit fuzzy, photographs of stars. As Bond recalled in a letter of 1857: "On these occasions a long exposure of one or two minutes was required before the plate was acted upon by the light, and in this interval the irregularities of the Munich clockwork were so large as to destroy the symmetry of the images, while the smaller stars of the second magnitude would not 'take' at all."[36] To create their photographic images, Bond and Whipple needed to master the delicate chemistry of the daguerreotype process, which involved treating a highly polished silver plate with nitric acid, iodine fumes, toxic mercury, and other chemicals. Because of the complexities, not to mention hazards, daguerreotypes were largely replaced with wet-plate (also called collodion) photography. In 1857, Cambridge telescope makers Alvan Clark and Sons added a new sky-tracking mechanism to the Great Refractor that allowed the telescope to smoothly track the movement of the stars as the earth rotated during the necessarily long exposures. On April 27, Bond and Whipple captured a clear image of the stars Mizar and Alcor.[37] It was one of the first clear pictures ever captured of stars.[38]

THE CAMBRIDGE U.S. EQUATOREAL.

The Great Refractor, Harvard College Observatory, lithograph by B. F. Nutting from a sketch by
A. Sonrel, 1848. *Source:* Harvard–Smithsonian Center for Astrophysics.

Alvan Clark and Sons

In the second half of the nineteenth century, Alvan Clark and Sons of Cambridge was the leading American manufacturer of telescope lenses, making high-quality lenses or full telescopes for dozens of colleges, universities, government agencies, and private citizens. Five times, the Clark firm made lenses for the largest refracting telescopes in the world. The Clark family's precision lenses were recognized as unsurpassed even by the best European firms. The family began making telescopes and scientific instruments around 1846, and the firm they founded continued to operate in Cambridge until 1933.

Prior to working on lenses, Alvan Clark had earned a living as a painter of portraits and miniatures in Boston, and he did not fully give this up until 1860, when his lens business finally generated enough income.[39] In 1860, he built a new manufacturing complex on Henry Street in Cambridgeport at the edge of the Charles River, with a workshop, observatory, and houses for himself and his two sons. The firm never employed more than a dozen workers, but its products were in high demand across the country. Alvan and his son Alvan Graham Clark ground sheets of European-made glass into telescope lenses called objectives. George Bassett Clark, Alvan's other son, made telescope barrels and the mechanical apparatus required to complete functioning telescopes. George was hailed as a gifted mechanic, while Alvan senior and junior made their reputation by hand-grinding lenses to compensate for optical irregularities in the glass sheets. "From the first," wrote historian Deborah Jean Warner, "Clark lenses were probably equal to any ever made."[40]

A late nineteenth-century arms race in astronomy meant even bigger and bigger telescopes, and the Clarks made the biggest. In 1860, the University of Mississippi ordered an 18.5-inch lens. Due to the outbreak of the Civil War, it ended up instead at the Dearborn Observatory in Chicago. The Clarks completed a 26-inch lens in 1873 for the US Naval Observatory, and in 1883, they built a 30-inch lens for an observatory in Pulkovo, Russia. In 1887, they built a 36-inch lens for the Lick Observatory in California and a decade later, a 40-inch lens for the Yerkes Observatory operated by the University of Chicago.[41]

Harvard's Great Refractor telescope had been made in Germany before Clark's firm had established its reputation. But the Clarks often worked for Harvard in the late nineteenth century, creating telescopes and astronomical equipment for the observatory and spectroscopes for Harvard chemists and physicists. The Clarks were themselves amateur astronomers and were able to make important observations, including finding a companion to the star Sirius in 1862 and discovering other binary stars—a pair of stars linked together in the same orbit—as they tested their newly polished lenses on the telescope mount installed at their factory.

The cover of *Scientific American*, September 24, 1887. The illustration contains four views of the Alvan Clark factory at the intersection of Brookline and Henry streets in the Cambridgeport neighborhood of Cambridge. *Source:* Courtesy, Linda Hall Library of Science, Engineering, and Technology.

The Harvard Computers

Edward Charles Pickering, a Harvard graduate and MIT physics professor, was named director of the Harvard Observatory in 1877 at the age of thirty-one; he was to hold the position for forty-two years. Shortly after assuming the directorship, Pickering began an ambitious project to photograph all of the stars in the sky, a project that resulted in the 1903 publication of a "Photographic Map of the Entire Sky." Pickering's goal was to document the position, brightness, and color of every star that could be observed telescopically to serve as a resource for the world's astronomers.

Photographs were not yet widely used in astronomical research in the 1880s. Instead, most astronomers relied on their own eyes to make visual observations through telescopes, recording these observations by hand on paper. Using long-exposure photography, however, Pickering could record images of stars too dim to be seen by the naked eye. And he could separate the act of observing through a telescope (at night in an unheated observatory dome) from the analysis and recording of astronomical data, which could be performed during daytime hours in a heated room by a separate staff of "computers."

To perform the analysis and recording work, Pickering began to hire women, who could be paid less than comparably trained men, thus stretching his limited funds. In 1881, Williamina Fleming, a Scottish woman who had emigrated to the United States, became a permanent member of Pickering's staff. Within a few years, Fleming began to compute the magnitudes and locations of stars from photographic plates, and she would go on to become a celebrated astronomer in her own right, making important contributions to the field at the Harvard Observatory for nearly thirty years.[42]

In 1886, Pickering convinced a wealthy New York widow, Anna Palmer Draper, to fund the creation of a catalog of the stars by capturing spectroscopic images of them.[43] Spectroscopy uses prisms to break up the light from an object into its constituent wavelengths, which appears as a spectrum from red to violet. By comparing the spectral patterns of light coming from distant stars to those created in a laboratory by burning elements like hydrogen and helium, astronomers can confirm the chemical composition of stars.

With Draper's financial commitment, Pickering hired more staff, many female, to analyze the photographic plates created at Harvard's observing stations in both the Northern and Southern hemispheres. Job opportunities were nearly nonexistent for a late nineteenth-century woman with an interest in the physical sciences, and the Harvard Observatory was able to attract an extraordinarily talented staff. The women became known as the Harvard Computers. Observatory staff created and studied more

Edward Pickering and Williamina Fleming supervising the women computers at the Harvard College Observatory, 1891. *Source:* Courtesy, Harvard College Observatory.

than 500,000 plates of the heavens between the 1880s and 1993.[44] The Henry Draper Catalogue, containing the spectroscopic signatures of more than 225,000 stars, was published by the Observatory between 1918 and 1924. The catalog and the star classification scheme it pioneered remain resources for astronomers today.[45]

Pickering's primary goal was to catalog the stars rather than advance particular theories about the heavens. Nonetheless, he and his staff made important discoveries. Using spectroscopic data, Pickering discovered a binary star in 1889; Antonia Maury, one of Pickering's recent hires, discovered a second binary pair later that year. In 1893, Williamina Fleming discovered a nova—a newly born star—from photographic plates made at Harvard's Southern Hemisphere observatory at Arequipa, Peru, and discovered two more novas in 1895. With Pickering's support, these women and others received credit in academic publications for their efforts, which was very unusual at the time. Between 1893 and 1908, one notable computer, Henrietta Swan Leavitt, discovered

Photographic plate from the Harvard College Observatory's glass plate collection.
Source: Courtesy, Harvard College Observatory.

Photographic plate from the Harvard College Observatory's glass plate collection, with marks made by astronomer Henrietta Swan Leavitt. *Source:* Courtesy, Harvard College Observatory.

more than 1,700 variable stars—stars whose brightness varies over a period of time rather than remaining constant. In 1912, Leavitt established an important relationship between the brightness of variable stars and the time period of their fluctuation, known to astronomers as Leavitt's law.[46]

Over the course of his career, Pickering "transformed a small, poorly endowed research center engaged chiefly in the visual observation of celestial objects into a large, well-endowed, well-equipped institution that fully exploited the newest methods of spectroscopy and photography," according to historians Bessie Zaban Jones and Lyle Gifford Boyd.[47] "Pickering, with his talented and devoted assistants . . . had established a uniform system for measuring the brightness of stars, and produced fundamental catalogues of stellar magnitudes. They amassed a photographic library containing a history of the stellar universe over a continuous period of more than 30 years."

Pickering encouraged and supported a generation of astronomers, male and female, who led the field well into the twentieth century. One example is Cecilia Payne-Gaposchkin, who in 1925 was the first person to receive a PhD in astronomy (rather than physics) at Harvard. In her dissertation, Payne argued that the atmosphere of stars was made of hydrogen—a controversial theory at the time that was later extended to include the whole universe, says Jonathan McDowell.[48]

The Harvard-Smithsonian Center for Astrophysics, still based on Observatory Hill, has in recent years employed more than nine hundred people. In 2005, the center undertook an initiative to digitize the more than 500,000 glass plates in its collection—the legacy of the Bonds and the Clarks, as well as Edward Pickering and the Harvard Computers. The images, about half of which were digitized by 2018, serve as a record of what has transpired in the sky over the course of a century.

ASA GRAY: BOTANY AND NATURAL SELECTION

On September 7, 1857, the English naturalist Charles Darwin wrote a letter to his friend, Harvard botanist Asa Gray, outlining in print for the first time his theory of evolution by natural selection. Darwin wrote: "As you seem interested in subject, & as it is an immense advantage to me to write to you & to hear ever so briefly, what you think, I will enclose . . . the briefest abstract of my notions on the means by which nature makes her species."[49] In his abstract, Darwin reasoned:

> I cannot doubt that during millions of generations individuals of a species will be born with some slight variation profitable to some part of its economy; such will have a better chance of surviving, propagating, which again will be slowly increased by the accumulative action of Natural Selection; and the variety thus formed will either coexist with, or more commonly, will exterminate its parent form.

Darwin, a trained biologist but not a specialist in botany, lacked Gray's detailed knowledge of the plant kingdom and needed Gray's help to prove his hypothesis. In the mid-nineteenth century, Gray was a pioneer in describing the geographical distribution of plant species, based on specimens that a network of collectors sent him from Europe and Asia and from expeditions in the newly explored Western United States.

In November 1859, Darwin published *On the Origin of Species*, detailing his novel theory and acknowledging Gray's contribution.[50] Gray received an early copy, perhaps the first copy in North America, and arranged in 1860 to have it republished in the United States.

Botanist Asa Gray. *Source:* Courtesy, Harvard University Herbaria.

The two men first met in London in 1839 and again in the spring of 1851 when Gray went on a year-long visit to Europe.[51] By 1855, they were regular correspondents, with Darwin often asking Gray about the geographical range of plant species included in Gray's previously published work.[52] Gray had noted similarities in the plant species of Eastern North America and those of Japan, which attracted Darwin's attention, and Gray undertook statistical analysis of these patterns at Darwin's behest. Gray advanced the idea that the two sets of species had emerged from a single ancient geographic location but had been separated by the emergence of glaciers. Gray's analysis supported Darwin's nascent theory of natural selection—the idea that the development of species is driven by random mutations that help organisms survive in their natural environment and reproduce.

As an 1890 biography noted, "Gray's comprehensive knowledge of the plants of the world, of their distribution, and specifically of the relations of North American species, genera, and orders to those of the other continents, and the precision of his knowledge, enabled him to be of much service to Darwin in the preparation of the first edition of the *Origin of Species*."[53] Gray became a close friend and regular correspondent of Darwin's, visiting his home in Kent, England, in 1868 and 1881 and serving as the chief American defender of his ideas.

Gray, a professor at Harvard from 1842 to 1872 and director of the University's botanic gardens, was the most influential botanist in nineteenth-century America. Guided by an ambition to catalog all of the plants of North America, he became a leader in plant taxonomy: the describing, identifying, naming, and classifying of botanical species.

Asa Gray in his office on Garden Street in Cambridge. *Source:* Courtesy, Harvard University Herbaria.

A dried specimen of *Calla palustris* (bog arum) from the Asa Gray Herbarium, collected in Massachusetts in 1870. *Source:* Courtesy, Harvard University Herbaria.

Gray was raised near Utica, New York, and trained as a doctor, graduating from medical school in 1831 at the age of twenty. His real passion, he discovered, was plants, so he studied botany for several years. Jobs as an academic botanist were not easy to find, but in 1838, Gray was named professor of botany and zoology at the newly founded University of Michigan. He was instrumental in collecting nearly 3,700 books for its library during a year-long research trip to Europe, but he never ended up teaching at Michigan due to financial difficulties there. In March 1842, he was offered a professorship at Harvard.

In 1844, Gray moved into a stately Federal-style house on Garden Street in Cambridge. The home occupied the site of the Harvard Botanic Garden, which had been established in 1805 for the cultivation of medicinal plants but had languished prior to Gray's arrival. Gray's move to Garden Street improved both his living situation—he had previously lived in rented quarters on Mason Street—and his botanical research.[54] He intended to collect North American plants to grow in the garden and to trade plants with his European contacts.[55] In addition to tending the garden, Gray built a library of dried plants. Sorting, analyzing, and preparing these specimens for preservation was no small task, and it occupied Gray for decades. In 1864, Gray donated his collection to Harvard in return for the promise that a suitable fireproof building would be erected to house them. Today, the collection, located on Divinity Avenue in Cambridge, contains nearly two million specimens.[56]

On moving to Harvard, Gray met and befriended many of his neighbors, including poet Henry Wadsworth Longfellow, astronomer William Cranch Bond, and fellow naturalist Louis Agassiz. There were not many professional scientists in 1840s America, let alone full-time botanists. Science was still largely a gentleman's hobby, without many institutional sources of support. Gray helped to establish a scientific club in Cambridge with other Harvard faculty and devoted amateurs, where they discussed and debated the scientific issues of the day, including Darwinian evolution.[57]

The middle decades of the nineteenth century were an era of great westward expansion in America: travel by covered wagon and scattered railroad lines was followed in 1869 by the completion of a transcontinental railroad. In the 1840s and 1850s, Gray developed a network of plant collectors who sent him materials from locations as far away as Oregon, Texas, and northern Mexico. He was inundated with plants from his eager collectors, which he intended to carefully study and catalog, in the process identifying the geographic ranges of individual species. As biographer A. Hunter Dupree remarked: "Any humble clergyman or schoolteacher in Alabama or Illinois considered it his inalienable right to collect some plants in his neighborhood, dry them, send them to Professor Gray of Cambridge—possibly the only botanist's name he had ever heard—and expect a letter by return mail which contained the names of his plants and

Asa Gray House and Botanical Garden on Garden Street in Cambridge. *Source:* Courtesy of the Patsy Bowdoin Collection, Cambridge Historical Commission.

words of encouragement and praise. More often than not, he got the letter."[58] Plant collections spilled out of the space in Gray's home that he had dedicated to his herbarium: "finally the whole house was crammed with plants—plants in the dining-room, in the attic, in the closets, and in the bedrooms."[59]

As a down payment on his promise to survey the entire plant kingdom of the North American continent, in 1848 Gray published *Manual of the Botany of the Northern United States, from New England to Wisconsin and South to Ohio and Pennsylvania Inclusive*. In 1857, he published a text for high school students titled *First Lessons in Botany and Vegetable Physiology*. An even more elementary work, called *How Plants Grow*, was published a year or so later.[60] The first volume of his major work, the *Synoptical Flora of North America*, was published in 1878, and a second volume appeared in 1884; together they comprised almost a thousand pages of small type.[61] He was, at heart, more a cataloguer than philosopher; despite his productivity as a published author, no great theories of evolution or plant genetics emerged from his pen.

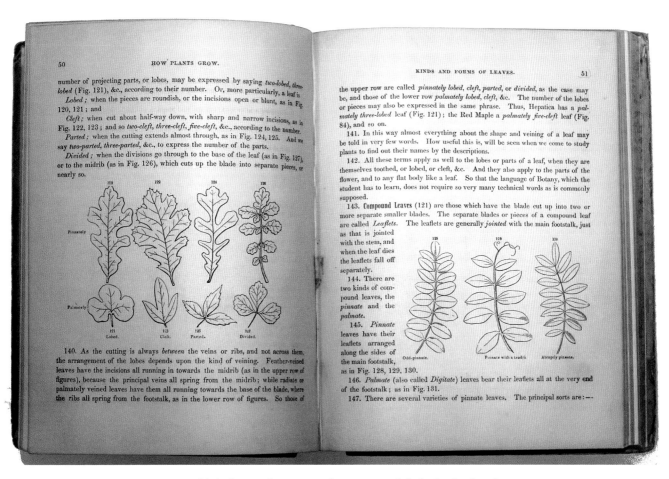

In 1858, Asa Gray published *How Plants Grow*, a botany manual for high school students.
Source: Photo by the authors.

Gray believed in closely observing the natural world and drawing scientific conclusions from those observations rather than from preformed theories or intuitions. In his empiricism, Gray distinguished himself from Louis Agassiz, his more famous Harvard colleague. Agassiz sought to find evidence of the hand of God in the natural world and to prove the divine origins of animal and plant species but increasingly strayed from observational science in his quest.[62] Gray was not the showman and orator that Agassiz was, and he maintained a much lower public profile, but he was less blinded by prejudices and more committed to the scientific method. Gray, though a devout Christian, saw no need for mystical explanations of the evolution of the plants he studied.[63]

Through public debates, conversations with his Harvard colleagues, and his published writing, Gray fiercely defended the Darwinian revolution and undermined Agassiz's clout. By the late 1860s, Gray's reputation as a scientist began to eclipse Agassiz's, particularly in the scientific communities of Europe.[64]

An 1890 biography of Gray, written two years after his death, notes: "American botanical science, wrought out so largely in its details, its system, and its philosophical relations, by his labors, is his monument."[65]

EBEN HORSFORD AND BAKING POWDER

If you look in your kitchen cabinet right now, chances are pretty good that you'll be able to find a can of baking powder. It's so commonplace today that it's hard to imagine someone invented it, but modern baking powder was created in the 1850s by a colorful Harvard chemist named Eben Norton Horsford.

Today, baking powder is a mixture of three ingredients: baking soda, also known as sodium bicarbonate; a weak acid, such as monocalcium phosphate; and a filler such as corn starch. Each of the three ingredients plays an important role in the leavening process that allows cakes and quick breads to attain their characteristic light, fluffy texture. Baking soda, chemically speaking, is a base. Monocalcium phosphate is baking soda's opposite, an acid. When these acids and bases are combined in water, they interact chemically, resulting in the release of carbon dioxide, a gas. The carbon dioxide creates the pockets of air in a baked good, keeping it from becoming a dense brick of flour and sugar. The third ingredient, corn starch, is a desiccant; it absorbs moisture from the atmosphere, keeping the two main ingredients dry so they do not interact prematurely. This allows the resulting chemical reaction to be predictable, which means that bakers using the same amount of baking powder will achieve the same amount of rising action in every cake.

For millennia, bakers used yeast to generate leavening in their breads. Yeasts are microscopic fungi; even in the dried form used by bakers today they are living organisms. Yeast cells feed on starches and sugars such as those in wheat flour. As yeast cells digest their food, they release carbon dioxide, which makes bread dough rise. As a natural product, yeast can be unpredictable; a given weight of dried yeast may contain a mixture of inactive and living organisms or yeasts of different strains and potency. In the 1800s, yeasts were thought by some people to be agents of decay and even associated with moral failing.[66] Scientists began the search for more reliable leavening agents. In the 1840s, cream of tartar—a byproduct of wine making—was found to generate carbon dioxide when mixed with baking soda.[67] But in the nineteenth century, wine production was mostly limited to Europe, meaning that American bakers had to import cream of tartar at great expense.

Horsford began to study the problem in 1854.[68] He proposed a substitute for cream of tartar: monocalcium phosphate made by treating animal bones with sulfuric acid. Horsford received a patent in 1856 for the process of making monocalcium phosphate. To prevent a premature reaction between the two primary chemical agents in his baking powder, Horsford proposed adding corn starch so that the ingredients could be packaged together, mixed in the correct proportions in a factory, and shipped ready for use by bakers. Horsford—who held the title of Rumford Professor and Lecturer on the Application of Science to the Useful Arts at Harvard—established the Rumford Chemical Works in East Providence, Rhode Island, where he manufactured the new product.[69] The American Chemical Society considers the invention of baking powder of such importance that in 2006 it named the development of baking powder a National Historic Chemical Landmark.[70]

In 1861, Horsford published *The Theory and Art of Bread Making: A New Process without the Use of Ferment*.[71] In the tract, he promoted the use of his chemical leavening over yeast. Horsford argued that the nutritional value of white bread was reduced by the action of yeast and the removal of phosphate-rich wheat bran. He noted that the chemical leavening he had invented worked more quickly than yeast (which often requires hours of rising time) and that baking powder required less skill from the home baker than recipes using yeast. He believed that the only purpose of yeast in bread making is to develop the air bubbles that permit easy human digestion of wheat.

Today, with the abundance of industrially produced foods available in America, Horsford's concern for the humble loaf of bread may seem quaint. However, in the nineteenth century, hunger was widespread, and inadequate nutrition even more common. As with so many areas of scientific investigation, food science promised to improve the health, robustness, and productivity of humankind.

Horsford was a somewhat unlikely hero in this arena. A New York native, he studied civil engineering, not agriculture or chemistry, at the Rensselaer School (now Rensselaer Polytechnic Institute) in Troy, New York. His interest in nutrition came during two years of advanced study at the University of Giessen in Germany. Horsford accepted the Rumford Professorship at Harvard in 1847 and taught chemistry there for sixteen years.[72]

Horsford had wide-ranging interests as a chemist and was a prolific author of scientific papers and reports. He analyzed grains and vegetables for their nutritional content. During the Civil War, he consulted with the US Army on efficient ways to feed soldiers in battle and developed a ration of compressed beef and parched wheat grits;

Portrait of Professor Eben Horsford. *Source:* Attribution 4.0 International (CC BY 4.0), Wellcome Collection.

Rumford Baking Powder tin. *Source:* Photo by the authors.

half a million packages were prepared per Ulysses S. Grant's orders.[73] Horsford advised the city of Boston on the best metal to use in its water distribution pipes. For the Massachusetts State Board of Health, he investigated the source of offensive odors near a meat packing plant in East Cambridge. He is credited as the inventor of condensed milk, which he reportedly created for use on the Arctic expeditions of explorer Elisha Kane. One of Horsford's assistants sold the process for condensing milk to the Borden Company, which continues to make it. All told, Horsford received thirty patents for his discoveries.[74] Leaving Harvard in 1863 due to the demands placed on him by his business interests, he spent the remainder of his career commercializing his discoveries and became a wealthy man.

In later life, Horsford became convinced that a group of Norsemen led by Leif Erikson had established a walled city in Watertown, Massachusetts, in the eleventh century. In 1887, he delivered an address at Faneuil Hall in Boston trumpeting the "Discovery of America by Northmen" and dedicating a statue of Erikson. His argument rested on his interpretation of early maps of the North American coast and his assertion that

Marker on Memorial Drive in Cambridge, placed by Eben Horsford to commemorate the purported landing of Leif Erikson on the banks of the Charles River in the year 1000. *Source:* Photo by the authors.

Trade card for Horsford's Self-Raising Bread Preparation, Rumford Chemical Works.
Source: Collection of the authors.

several place names in Massachusetts were of Norse origin. In 1889, he published *The Discovery of the Ancient City of Norumbega*, followed two years later by *The Defences of Norumbega*, based on archaeological investigations in Watertown. In the 1889 work, Horsford grandly announced: "I have to-day the honor of announcing to you the discovery of Vinland, including the Landfall of Leif Erikson and the Site of his Houses. I have also to announce to you the discovery of the site of the ancient City of Norumbega."[75] A review of the 1891 book in *Science* magazine noted: "He believes that he has discovered its stone-built walls, its ancient stone-paved streets, and the remains of its docks and wharves. Other local antiquaries see in these remains merely the vestiges of some dams, drains, and stone fences of the early New England farmers, and it appears that Professor Horsford has not succeeded in persuading any of the resident investigators of the interpretation he has so much at heart."[76] Nonetheless, in 1892 Horsford was knighted by the king of Denmark for his archaeological research on the Norse people.

Subsequent research has not supported Horsford's claims of Viking settlements in New England. But evidence of his conviction endures. On a lonely spot at the intersection of Memorial Drive and Gerry's Landing Road in Cambridge, behind Mount

Auburn Hospital, a stone monument reads "On This Spot in the Year 1000 Leif Erikson Built His House in Vinland." Erection of the marker was funded by Horsford and his baking powder profits. Putting his wealth to perhaps more fruitful use, he became a major benefactor of Wellesley College, endowing its library, establishing a fund for scientific instruments, and creating a pension fund.

An obituary in the *Proceedings of the American Academy of Arts and Sciences* notes that "After his retirement from his professorship, he continued to live in Cambridge until his death, as it had become endeared to him by his long residence and the brilliant society in which it then rejoiced."[77] The stately home he built on Craigie Street in Cambridge in 1854 remains a private residence.

LOUIS AGASSIZ: PUBLIC SCIENTIST

Louis Agassiz was the most famous scientist in America for nearly three decades, from his arrival in Boston in 1846 to his death in 1873. A compelling public speaker, he increased Americans' appreciation of science and argued persuasively for government funding of scientific research in the United States. At Harvard, where he was a professor of the natural sciences, he built an impressive collection of animal specimens and fossils, which became the Museum of Comparative Zoology in 1859. From his home on Quincy Street in Cambridge, Agassiz wielded enormous power in American science through advocacy, personal charm, and an enviable work ethic. He received—and thrived on—the admiration of politicians, farmers, industrialists, mechanics, and shopkeepers.

Pioneering Harvard philosopher and psychologist William James, who traveled on a collecting expedition with Agassiz as a young man, attested to the elder scientist's magnetism and influence more than twenty years after his death. He "came before one with such enthusiasm glowing in his countenance—such a persuasion radiating from his person that his projects were the sole things really fit to interest man as man—that he was absolutely irresistible. . . . He on the whole achieved the compass of his desires, studied the geology and fauna of a continent, trained a generation of zoologists, founded one of the chief museums of the world, gave a new impulse to scientific education in America, and died the idol of the public, as well as of his circle of immediate pupils and friends"[78]

But Agassiz also held views that, in today's world, are both widely rejected and unscientific. In his lectures and writings, Agassiz promoted the idea that people of African descent were intellectually inferior to Europeans. In the 1850s and 1860s, he rejected the emerging theory of evolution by natural selection for a much less scientifically fruitful idea: the notion that all living species emerged simultaneously from the hand of a divine creator. It is a challenge to balance a consideration of Agassiz's contributions to the natural sciences with his profoundly reactionary views about race and evolution but equally hard to ignore his impact on nineteenth-century America.

Agassiz the Naturalist

Agassiz was at heart a close observer of nature. He stressed the importance of looking at an animal, a vegetable, or a mineral in its native habitat where possible and describing it carefully and methodically. He encouraged comparisons between similar objects as a way to build knowledge. And he amassed great collections of items from the natural world to serve as his textbooks and laboratories. In this sense, he brought to America a passion for the scientific methods he had learned in Switzerland, where he was born; Germany, where he completed his advanced education; and France, where he launched his career as a geologist and zoologist.

William James, who traveled with Agassiz to Brazil in 1865, wrote about Agassiz's passion for direct observation: "'Go to Nature; take the facts into your own hands; look and see for yourself!' These were the maxims which Agassiz preached wherever he went, and their effect on pedagogy was electric."[79]

Agassiz established his scientific credentials in Paris in the 1830s, working under the noted naturalists Georges Cuvier and Alexander von Humboldt. His promise as a scientist came to attention through his *Recherches sur les poissons fossiles*, five volumes published from 1833 to 1843, in which he described and analyzed seventeen hundred different fish species from their fossilized remains.[80]

Following this achievement, Agassiz was one of the first naturalists to promote the then-novel idea that the continents had once been covered by great sheets of ice, which slowly receded as the earth warmed. Agassiz delivered a paper in Switzerland 1837 supporting the claims of other researchers that certain landforms and geological debris fields were evidence of glacial activity in the distant past.[81] Agassiz reiterated these ideas about the Ice Age in two books he published in the 1840s, based on his careful investigation of rock faces and landforms in the Alps and other areas of northern Europe. He saw the ancient glaciers as "God's great plow," powerful evidence of divine intervention in earthly affairs.[82]

Agassiz Comes to America

Agassiz was invited to Boston in 1846 to deliver the annual Lowell Institute lectures. His journey, part of a planned two-year research expedition in the United States, was funded by the Prussian government. He never returned to Europe.[83] The Lowell lectures proved wildly popular. The following year, he was named a professor of zoology and geology at Harvard's newly established Lawrence Scientific School. Agassiz's appointment represents the "beginning of the professionalization of American science," according to author Louis Menand.[84] For Agassiz, America was a vast and largely untapped trove of scientific information that he could study ahead of his European peers. In his first decade in the United States, Agassiz led explorations in the Northeast, the South, the Mississippi Valley, and the Great Lakes, collecting animal and fossil specimens that were sent back to Cambridge. He courted wealthy Boston industrialists to fund an encyclopedic survey of the natural history of the American continent, and he solicited contributions of specimens from collectors across the country. A ten-volume set titled *Contributions to the Natural History of the United States* was planned, to be issued by Boston publishing house Little, Brown and Company, though Agassiz completed only four of the volumes.[85]

In 1855, Agassiz wrote to a friend in Switzerland: "I have now been eight years in America, have learned the advantages of my position here, and have begun undertakings which are not yet brought to a conclusion. I am also aware how wide an influence I already exert upon this land of the future, an influence which gains in extent and intensity every year."[86]

Promoter of Science

Both in Boston and during his expeditionary travels in America, Agassiz became an evangelist of science, delivering lectures on the natural world to any audience that would hear him, including "lyceums, teachers' associations, and farmers' institutes."[87] His lectures were often accompanied by elaborate illustrations he drew on a blackboard as the audience watched in amazement.

In 1853, Agassiz began encouraging Americans to send him dead fish in the mail. He distributed thousands of circulars on the proper means of collecting and shipping the specimens he intended to use in creating a catalog of the fishes of the United States. Properly deputized by Agassiz, collectors across the country complied with his request, eagerly joining his family of amateur scientists.[88] The fish piled up in Cambridge, awaiting the permanent home that was to become Agassiz's lasting monument, the Museum of Comparative Zoology, which opened in November 1860. The museum was the result of private donations and an appropriation from the Massachusetts legislature.

Agassiz was instrumental in encouraging the US Congress to establish the National Academy of Sciences (NAS) in 1863, creating a formal link between the federal

Louis Agassiz, 1862. *Source:* Courtesy of the Cambridge Historical Society.

government and the nation's scientists. The NAS—originally comprising fifty members, many from Cambridge—was charged with providing "independent, objective advice to the nation on matters related to science and technology."[89]

An obituary published in *Scientific American* in 1874 notes that Agassiz "seemed as if he were every one's immediate friend; his personality was of that magnetic order which appeals directly to the heart, and it as the charming simplicity of his manner, coupled with the glow of enthusiasm which pervaded his every utterance, that made even the dullest units of his vast audiences feel that the subject under treatment, though never so dry, was invested with new attributes of rare and before unseen interest."[90] There was no one else in Cambridge quite like Agassiz: a sophisticated European, man of the world, bon vivant, charmer. In Europe he might have been just another academic, but in the comparatively small pond that was Boston society, he was a big fish. He realized that there was growing wealth in America, a result of the industrial fortunes being made in businesses like textile production, shipping, and finance. He aimed to fill a void in America's young scientific establishment by tapping into that wealth.

Educator of Young Women

In 1855, Agassiz and his wife Elizabeth Cary Agassiz began operating a school for girls in their Quincy Street home. Their pupils would include the daughters of Ralph Waldo Emerson and a young Melusina Fay, the future cooperative housekeeping pioneer.[91] Elizabeth Cary Agassiz went on to cofound Radcliffe College, offering collegiate instruction to young women, and served as its first president. In their home school, Louis and Elizabeth Agassiz insisted on high standards of scientific inquiry and observation from nature. Unusual for his time, Louis Agassiz also offered women the opportunity to study at the summer school in natural history he had established at Pekinese Island off the coast of Massachusetts.[92]

Racial Views

In 1853, Agassiz traveled to Mobile, Alabama, to deliver a series of public lectures. In one, he claimed that "we see in the races a gradation parallel to the gradations of animals up to man," asserting that some races, as he perceived them, were inferior to others.[93] The following year, he penned an essay for a book entitled *Types of Mankind* in which he wrote that "it follows that what are called human races, down to their specialization as nations, are distinct primordial forms of the type of man." He argued, in essence, that God had created several different subspecies of humans and assigned

Museum of Comparative Zoology at Harvard University, 1875. *Source:* Collection of the authors.

them to the different continents. For anyone wanting to argue that Africans were not equal to Europeans and could therefore be enslaved without moral misgivings, Agassiz's claim served as a quasi-scientific justification. A follow-up, *Indigenous Races of the Earth*, was published in 1857, again with contributions from Agassiz. Louis Menand notes that Agassiz was not intending to defend the institution of slavery, which he saw as a violation of his religious beliefs. Agassiz insisted that he was simply a scientist following the evidence. Still, Agassiz wrote that human society would be better off if "we were guided by a full consciousness of the real differences existing between us and them, and to foster those dispositions that are eminently marked in them, rather than by treating them on terms of equality."

In the mid-nineteenth century, there was no societal issue more pressing, no conversation more critical to the future of the American republic, than the inherent rights of enslaved peoples of African descent. Agassiz wrote no memoir and left few written records outside his scientific publications and lecture notes, making it hard for later biographers to divine his inner thoughts. Whatever his personal views on the subject

Gravestone of Jean Louis Rodolphe and Elizabeth Cary Agassiz at Mount Auburn Cemetery, Cambridge and Watertown, MA. *Source:* Photo by the authors.

of race, Agassiz biographer Christoph Irmscher has argued, Agassiz wanted above all to prove his relevance to the national debate. As Irmscher notes, "Taking a position on race . . . assured Agassiz of the continuing interest of a non-academic audience, a priority that would grow in importance to him over the course of his American career."[94] Irmscher argues that Agassiz's racial beliefs were shared by many upper-class whites, in both the North and the South, but his attempt to justify his prejudices via science and his willingness to express them in his writings and lectures put him at an extreme end of the national conversation.[95]

Louis Agassiz remains a challenge to place in history—a great promoter of the important role that science could play in America, an early educator of women, yet also a deeply flawed racial theorist and a man who at times placed the interests of his own popularity over his devotion to science.

WALLACE SABINE AND ARCHITECTURAL ACOUSTICS

In the early months of 1896, if you had stood outside Harvard's Sanders Theatre after the last lecture of the day, you might have noticed young men lugging red velvet seat cushions out of the building, across Cambridge Street, and into the university's Fogg Museum. By dawn, the cushions would be back in their rightful place; another night of scientific research completed. The cushions were used to solve a problem plaguing a lecture hall inside the Fogg and would help lead to an entirely new field of science.

In 1895, the stately Fogg Museum had been recently completed, with an exterior of limestone and Ionic columns, designed by eminent New York architect Richard Morris Hunt. But students who slipped in to hear the fine arts lecturer Charles Eliot Norton couldn't make out what their professor was saying.[96] The acoustics in the new semicircular lecture hall were terrible. Harvard president Charles Eliot (a cousin of the fine arts professor) was so concerned that he personally approached a junior faculty member in the university's physics department for help. Colleagues told twenty-seven-year-old Wallace Sabine to ignore the request. It was a fool's errand, they said. But he took up the challenge and set out to figure out why Sanders Theatre was such a marvelous lecture hall and the Fogg so unsatisfactory, though the two sat directly across Cambridge Street from each other and had the same amphitheater-like shape.

Sabine set up a system to analyze the acoustical properties of the Fogg hall. From the speaker's platform, he would briefly switch on a specially modified organ pipe. His research assistants, standing at various spots in the room, would take note of the sound's reverberation time—the amount of time that elapsed until they could no longer hear the tone. A sound emitted at the Fogg's podium, they learned, bounced around the room much longer than it did in Sanders Theatre, explaining why it was so hard to understand. As Sabine wrote at the time:

> A word spoken in an ordinary tone of voice was audible for five-and-a-half seconds afterwards. During this time even a very deliberate speaker would have uttered twelve or fifteen succeeding syllables. Thus the successive enunciations blended into a loud sound, through which and above which it was necessary to hear and distinguish the orderly progression of the speech. Across the room this could not be done; even near the speaker it could be done only with an effort wearisome in the extreme. . . .[97]

In one of his laboratory notebooks, Sabine recorded his methodical tests of how "plaster walls, chairs, oil paintings, glass, green plants, and rugs" each affected the sound quality of the space.[98] Night after night, Sabine and his research assistants borrowed seat cushions from Sanders Theatre, running them over to the Fogg building to see how they changed the sound quality. They placed a few lengths of seat cushions into the Fogg lecture hall at a time, making acoustic measurements using the organ pipe, then adding more cushions and making new measurements. On many nights, Sabine and his assistants worked until dawn, sleeping during the daytime. As a reason for his

Wallace Sabine, ca. 1922. *Source:* Harvard University Archives.

Fogg Museum lecture hall, after 1895. *Source:* Courtesy, Harvard Art Museums.

unconventional work hours, he later explained that Cambridge "was quiet only after the street cars had stopped running late at night and before the milkcarts started rolling over the cobblestones in the morning."[99]

President Eliot became impatient with Sabine's lack of rapid conclusions about what was now certainly a major embarrassment to the university, but Sabine was doing more than fixing an isolated problem: he was inventing a whole new field of physics. With all of the Sanders cushions crowding the 436-seat Fogg auditorium, Sabine calculated that he could reduce the reverberation time of the room from 5.62 seconds to 1.14 seconds, a considerable improvement.[100] Finally, in September 1898, at Sabine's recommendation, Harvard installed blankets made of hair felt on nearly two dozen surfaces in the Fogg lecture room and covered the speaking platform with carpet.[101] Though improved, the building was never an acoustic success; later renamed Hunt Hall in honor of its architect, it was demolished in the early 1970s to make way for an undergraduate dormitory.

John William Strutt, a British scientist also known as Lord Rayleigh, had studied the physics of acoustics in the 1870s and 1880s, but his work didn't focus on architecture.[102]

Sabine brought a real-world approach to the science. "He started applying acoustical concepts and the scientific method to actual architectural space," says Benjamin Markham, the director of the architectural acoustics group at Acentech, a consulting company in Cambridge.[103] "The notion of acoustics in rooms wasn't new. By trial and error, [others] had lots of ideas about all this. But actually applying the scientific method to the thought of acoustics in rooms specifically—that was unique and that was what happened" under Sabine. That's why he's credited with founding the field, says Carl Rosenberg, Markham's colleague at Acentech. "I do feel that we're acolytes."[104]

Sabine's work at the Fogg lecture hall was only the beginning of his career in acoustics. In October 1898, Harvard's President Eliot asked Sabine to advise the architects of a new performance hall being planned for the Boston Symphony Orchestra. At that point, Sabine had lots of raw data but no formulas he could use to analyze the design of an unbuilt room. He spent several weeks feverishly calculating, trying to derive a mathematical equation for determining reverberation times. As twentieth-century acoustician Leo Beranek notes, "On Saturday night, October 29, [1898,] the answer came to him. His mother wrote in her diary, 'his face lighted with gratified satisfaction,' as he announced quietly, 'I have found it at last.'"[105] The next day, Sabine wrote to Eliot, accepting the Boston Symphony assignment. Rosenberg says Sabine "redirected [architects] McKim, Mead and White from what would have been a horrible thing."

For the Boston Symphony, Sabine worked with architects to perfect the acoustic qualities of the planned concert hall while increasing the number of seats. He devoted nearly two years—again, working mainly at night—to refine the symphony hall design.[106] He eliminated balconies at the side of the stage so that sound would carry further into the main hall and instead suggested two levels of balconies at the rear of the hall. He recommended an under-floor ventilation system to limit noise from the heating system. Boston's Symphony Hall opened in 1900 to significant acclaim and has maintained a reputation as one of the most acoustically refined venues ever built for orchestral music.

To translate his work into findings other researchers could replicate, Sabine needed to convert his basic unit of sound absorption—linear feet of red velvet Sanders Theatre cushions—into a more universal measure. He discovered that he could relate the sound absorbance of his cushions to that of an open window. Through experimentation, Sabine determined that the acoustic absorbance of an open window was related to its square area and that his cushions absorbed about 80 percent of the sound energy of an equivalently sized window opening.[107] Sabine's work involved months of painstaking measurements and tedious calculations. To test his nascent theories, he evaluated the acoustical properties of meeting and lecture rooms across Harvard's campus.

Postcard of the 1895 Fogg Museum of Art (later known as Hunt Hall) at Harvard University. The building was demolished in 1973 to make way for a dormitory complex. *Source:* Collection of the authors.

He published the results of his research in 1900 in the *American Architect and Building News*. In the publication, he reported that the reverberations in a lecture hall were independent of where the speaker stood and that the efficiency of sound-absorbent materials was likewise independent of their position in the room.[108]

Sabine established himself as an expert—for a time, the *only* American expert—in the fledgling discipline of building acoustics. His consulting engagements included work for the House of Representatives in Washington, DC, and hundreds of court houses, churches, and performance venues from Bangor, Maine, to Minneapolis, Minnesota.[109] Sabine pursued ways to describe, measure, and manipulate the acoustical qualities of each architectural space. According to Leo Beranek, Sabine "noted that different uses of a hall, for example, for speech as opposed to piano or chamber music, require different reverberation times."[110] In 1915, the Remington Typewriter Company of New York asked him to help it produce quieter typewriters.[111] Sabine even studied the sound absorption properties that people—the members of an audience—exhibit.[112]

Red velvet seat cushion from Harvard University's Sanders Theatre used by Wallace Sabine in his 1896 experiments, 2018. Per the printed explanation accompanying the cushion, the cushion consists of hair covered with canvas and light damask. *Source:* Collection of Acentech, Cambridge; photo by the authors.

Details about Sabine's character and motivations are limited. Even the author of a 1933 biography had to concede that "Wallace Sabine lived during his entire life behind a 'protective crust of reserve.'"[113] What we do know is this: Sabine was born in 1868 in Richwood, Ohio, forty miles north of Columbus. His father was a graduate of Harvard Law School, and his mother had studied with the naturalist Louis Agassiz in Cambridge.[114] Entering Ohio State University at the age of fourteen, two years younger than most freshmen at the time, Sabine graduated after four years with a concentration in physics. In 1886, he pursued a PhD at Harvard while his older sister enrolled at MIT. His mother accompanied her children to Cambridge, and the family boarded on Story Street for a year before moving to Dana Street.[115] As a graduate student at Harvard, Sabine was described as being quiet and deeply absorbed in his studies, keeping a small circle of close friends, and exhibiting a fearsome work ethic. In 1890, at the age of twenty-two, he was appointed an instructor in the physics department at Harvard, a fact that delighted his mother, and five years later he was named an assistant professor and began his career as an acoustics expert.[116]

Sabine quite literally worked himself to death during World War I. He traveled to France on scientific missions in 1916 and 1917, visiting the battlefront and advising

French and Italian armed forces.[117] In Europe, he fell ill with an infected kidney but refused to let sickness get in the way of his work. After returning, he shuttled between Cambridge and Washington, DC, balancing his duties at Harvard with consulting on the design of aircraft for the US military.[118] He died in January 1919 at the age of fifty-one of kidney cancer.

Sabine never sought fame. Nonetheless, as a measure of his importance to the field of acoustics, a unit of sound absorption—the sabin—was named for him, and the Acoustical Society of America still awards the Wallace Clement Sabine Medal to honor major achievements in acoustical science.

In Cambridge, Acentech engineers keep an original Sanders Theatre cushion in a clear acrylic case. Displayed on a bookshelf, the red velvet is dirty and worn, and the blue and white ticking on the back is threadbare. But to Rosenberg, the cushion remains an important artifact from the founding of his field.

Preparing to leave office in 1961, President Dwight D. Eisenhower warned Americans against the potential dangers of "the military-industrial complex." His speechwriters had at first used the phrase "the military-industrial-scientific complex"—but James Killian, MIT president from 1948 to 1959 and Eisenhower's special assistant for science and technology, convinced the president to drop the third part of the triad.[1] Still, the idea that academia belonged was embedded in his farewell speech. "In holding scientific research and discovery in respect, as we should, we must also be alert to the equal and opposite danger that public policy could itself become the captive of a scientific-technological elite," the World War II hero said three days before leaving office. Although little remarked on at the time, this speech, historian Douglas Brinkley argues, became a rallying cry for the antiwar movement that emerged later in the decade.

People and institutions in Cambridge have served as key players in creating this military-industrial-scientific triad but have also been instrumental in highlighting its dangers and consequences.

Cambridge's role in US military history began even before the founding of the nation. On July 2, 1775, General George Washington took command of the newly formed Continental Army in Cambridge, preparing to fight the British. That first mustering, legend has it, happened under an elm tree on the Cambridge Common, just north of what is now Harvard Square. Washington described his assembled troops as "a mixed multitude of people under very little discipline, order or government."[2] Drilling regularly on the Common, he reshaped them into a fighting force, based on his experience as an officer during the French and Indian War. Washington had never before led an entire army, but his success helped propel him to the presidency.

Loyalists to the British Crown who had built estates in Cambridge fled in the face of the gathering Patriot army. Washington moved his headquarters into the abandoned house of John Vassal, a grand Georgian mansion on what today is Brattle Street. (The house is now best known for its second most famous resident, poet Henry Wadsworth

Longfellow.) While living in the Vassal House, Washington assembled men and armaments and made plans to conduct extended warfare with the British, whose navy was stationed in Boston Harbor. In all, Washington spent nearly nine months in Cambridge. In March 1776, the Continental Army placed cannons on a hill just south of Boston, threatening British supply lines. Realizing that the city was indefensible, the British headed out to sea. Military action shifted south for the remainder of the war.

The tree where Washington first gathered his troops was nicknamed the Washington Elm in his honor. A Cambridge fixture for well over a century, the tree died in 1923, but its offspring continue to grow nearby.

The people of Cambridge played a less pivotal role in the Civil War, though ninety-seven local men formed Company C, 3rd Regiment, Massachusetts Volunteer Militia, which mustered in Central Square on April 17, 1861.[3] Each man took a horse from Pike's stable, which sat at the back of an alleyway now routinely spray-painted by local artists. The 136 Harvard students and alumni who died fighting for the Union cause were honored by the construction of the university's majestic Memorial Hall, completed in 1878.

Mobilizing for World War II

In early 1940, former MIT engineering dean **Vannevar Bush** presented a proposal to President Franklin Delano Roosevelt to create a council of leading scientists that would coordinate research efforts in support of American defense. That June, Roosevelt signed an executive order creating the National Defense Research Committee (NDRC) to coordinate and conduct scientific research on challenges facing the development, production, and use of weapons.[4] Bush chaired the committee; its other eight members included MIT President Karl Compton and Harvard President James Conant. The committee had its first meeting on July 2, 1940, a year and a half before the Japanese attack on Pearl Harbor that led the United States to a formal declaration of war.

By the end of 1940, engineers and scientists in Cambridge were busy developing militarily useful technologies like **microwave radar**. Soon, the city's factory workers were producing polarized goggles for soldiers, the candy packaged in their rations, and other wartime essentials like soap and bicycle tires.

In his Harvard chemistry lab, **Louis Fieser** began working on incendiary chemical cocktails that could cause explosive fires when dropped on enemy buildings. The form of jellied gasoline he called napalm was used extensively in fire-bombing Germany and Japan. Napalm came to greater public attention and controversy when it was used in the Vietnam War several decades later.

The Washington Elm on the Cambridge Common, ca. 1909. The tree, as legend has it, is where George Washington took command of the Continental Army in 1775. The tree died in 1923.
Source: Detroit Publishing Company photograph collection, Library of Congress.

The links between the US military, private industry, and academia forged partly in Cambridge during World War II proved durable. Raytheon, founded before the war by Vannevar Bush, became a major developer of weapons systems. **Harold Edgerton**, an MIT professor and strobe photography pioneer, had a role in the development and testing of nuclear weapons during the Cold War. **Charles Stark Draper's laboratory**, which spun off from MIT during the Vietnam War, developed some of the earliest navigation systems for airplanes in the 1930s and became a key contributor to both the Apollo moon missions and American guided missile technology.

Government funding of scientific research at MIT and Harvard spurred many high-technology innovations after the war, particularly in the fields of electronics, computing, physics, and biology.

Some scientists who joined the MIT Radiation Laboratory during the war remained in Cambridge afterward. Physicist Edward Purcell returned to the Harvard faculty, where, in 1946, he demonstrated for the first time the principle of nuclear magnetic resonance (NMR), a property of atomic nuclei placed in strong magnetic fields. Today, we mainly know NMR for its use in a medical imaging technique—magnetic resonance imaging (MRI)—but the principle also has important uses in spectroscopy, astronomy, and other fields.

Ringing the Alarm Bells

While many Cambridge academics were instrumental in creating the military-industrial-scientific complex, others were among the first to highlight the dangers of an increasingly militarized American society.

Microbial geneticist Salvador Luria, who fled Nazi-controlled Paris by bicycle during World War II, became a vocal critic of nuclear weapons testing during the 1950s and joined the MIT faculty in 1959. He transformed the Institute's small biology department into a powerhouse in biological research and was awarded a Nobel Prize for his scientific work. Luria was actively opposed to the Vietnam War, often participating in letter-writing efforts and helping to buy full-page newspaper ads.[5]

Luria's most vocal MIT colleague and friend, **Noam Chomsky**, arguably became as famous for his political activism as for his pioneering work in the field of linguistics. For more than six decades, Chomsky has critiqued American economic and foreign policy, calling political, academic, and media figures to task for what he has argued is their hypocrisy and subservience to moneyed interests.

In 1969, at the height of national debate about America's role in Vietnam, MIT students and faculty founded the Union of Concerned Scientists to protest the use of science in advancing the war effort and to redirect academic research toward pressing social and environmental issues. A half century later, from its Cambridge headquarters the Union continues to advocate for the use of science in creating a healthy planet and a conflict-free world.

Molecular biologist Matthew Meselson, who joined the Harvard faculty in 1960, is known for his contributions in the field of genetics (see chapter 7) but has also helped to protect the world from the spread of biological and chemical weapons.

The people profiled in this chapter acted in good faith to advance the goals of a just world, though their perspectives and methods varied widely. All of them leveraged the unique resources they found in Cambridge in advancing their objectives.

VANNEVAR BUSH: SCIENCE IN THE NATIONAL INTEREST

On April 3, 1944, with the outcome of World War II still uncertain, the cover of *Time* magazine featured a drawing of a man named Vannevar Bush with a vacuum tube. The magazine proclaimed him the "General of Physics" and described the organization he ran, the Office of Scientific Research and Development (OSRD), as the fifth branch of the military—as crucial to the war effort as the army, navy, air force, or marine corps. Hammering his desk with his fist for emphasis, Bush told the *Time* interviewer that "if we had been on our toes in war technology ten years ago, we would probably not have had this damn war."

The contributions of America's scientific community to its World War II victory have not always been heralded, with the exception of the Manhattan Project that developed the atomic bomb. But scientists were integral to the fight. The man most directly in charge of these efforts, Vannevar Bush, had spent two decades at MIT—as student, professor, administrator, entrepreneur, and a key figure in the development of computers. At the OSRD, he drew on all of his Boston-area experience in directing the six thousand scientists, engineers, and support staff at three hundred different institutions who worked under him. Reporting directly to President Franklin Delano Roosevelt, Bush oversaw all nonmilitary research related to the war, in fields as varied as microwave radar, antibiotics, and artillery fuzes. The US Army's massive Manhattan Project fell under his purview. When Roosevelt died in office and Harry S. Truman assumed the presidency, Bush was the person who first briefed Truman in detail on the atomic bomb.

A year after the *Time* article, as World War II was coming to a close, Bush laid out a vision for how scientific research should be carried out in the United States. In a paper entitled "Science, the Endless Frontier," Bush described the role he believed basic science research should play in a democracy and the need for the federal government to

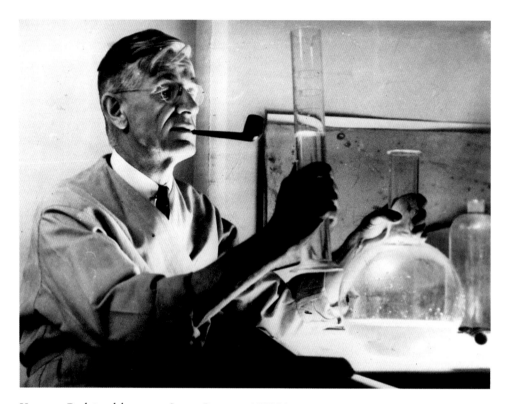

Vannevar Bush in a laboratory. *Source:* Courtesy, MIT Museum.

fund it. Before the war, research in the service of US national security had been conducted primarily by the military. But Bush argued that even when it was serving in the nation's defense, science should be carried out at universities.

The essay provided a roadmap the country has followed for more than seventy-five years. Bush's vision created "a unique set of capabilities unmatched around the world. . . . It transformed education in this country," said Robert Conn, president of the California-based Kavli Foundation, which supports basic research.[6] The scientific enterprise championed by Bush led to a flood of discoveries that drove the US economy in the postwar decades and made it a global leader in high technology.

Bush encouraged the creation of the National Science Foundation (NSF), a federal agency that funds research at American universities. In 2020, the NSF's annual budget topped $7 billion. In helping to establish an alphabet soup of important federal agencies, Bush exerted more influence on the nation's science policy in the postwar era than any other American, journalist Robert Buderi argues.[7]

Born in Everett, Massachusetts, and raised in the nearby town of Chelsea, Bush attended Tufts College (now University) and then earned a doctorate in electrical engineering from MIT in 1916. To supplement his income as a Tufts professor, Bush began running the research lab at a local company, the American Radio and Research Corp.[8] During World War I, he worked on antisubmarine technology for the US Navy— technology that was never used effectively. His frustration "forced into my mind pretty solidly the complete lack of proper liaison between the military and the civilian in the development of weapons in time of war, and what that lack meant," he would later recall.[9]

Bush moved to the MIT faculty in 1919. His experience in both industry and academia taught him the importance of crossing traditional boundaries. He founded the American Appliance Company, which, after a brief, unsuccessful venture in the refrigerator business, developed vacuum tubes for home radio sets. The company, renamed the Raytheon Manufacturing Company in 1925, was a commercial success, and its radio tubes made Bush a rich man.[10] Raytheon, based in Waltham, Massachusetts, became a major US aerospace and defense contractor.

Bush helped transform MIT from a polytechnic college that trained engineers for service in industry into a major scientific research organization.[11] In 1930, Karl Compton, the newly appointed MIT president, began to reorganize the Institute and two years later named Bush a vice president and dean of its school of engineering. Bush appointed research-oriented engineers to leadership positions and sought partners in industry who would advance his research-centered agenda. Compton's and Bush's reforms helped prepare MIT for the role it would later play during World War II.

Bush was an innovator in the field of computing in its early days. At MIT, he oversaw the creation of a differential analyzer, a room-sized mechanical device that could be programmed to perform mathematical calculations. To operate the complex machine, he hired a mathematician and engineer named Claude Shannon, who drew on his work under Bush when he later defined fundamental concepts of digital communications (see chapter 6).

Bush's ambition took him to Washington, DC, in 1938, where he accepted the presidency of the Carnegie Institution, a private philanthropy that funded scientific research. From his platform at Carnegie, Bush became increasingly influential in setting the nation's science and technology policy. The same year, he was appointed chair of the National Advisory Committee for Aeronautics, a precursor to the National Aeronautics and Space Administration (NASA).

Bush was 5'10" tall and trim, favoring dark three-piece suits and wide-rimmed spectacles.[12] He puffed on a pipe he carved himself. His wit earned comparisons to

A detail of Vannevar Bush's differential analyzer. *Source:* Courtesy, MIT Museum.

Vannevar Bush appearing as a guest on NBC television's *Meet the Press*, May 24, 1959.
Source: Courtesy, MIT Museum.

the Oklahoma-born actor and humorist Will Rogers. But Bush could also be an arrogant, imperious bully who harbored "a relentless, perhaps insatiable drive for power," according to biographer G. Pascal Zachary. Bush never avoided a fight. "My whole philosophy . . . is very simple," he told a few generals during World War II. "If I have any doubt as to whether I am supposed to do a job or not, I do it, and if someone socks me, I lay off."[13]

Bush remained in Washington until 1955, when he retired from the Carnegie Institution. He renewed his connection to the Boston area, retiring to suburban Belmont, Massachusetts, and serving on the MIT Corporation, the Institute's board of trustees. In his 1970 autobiography, Bush outlines all that he saw wrong with his country at the time: racial tension, student riots, pollution, an "absurd" war in Vietnam, and widespread poverty.[14] But he admonishes his readers to see the humor in life and to take faith from the progress that had been made—in part, through science. "The war effort taught us the power of adequately supported research for our comfort, our security, our prosperity," he writes. "And in the decades of peace that followed, we carried on, doing foolish things at times, with some waste, but on the whole, wisely . . . and that fact is something that we can be proud of."

THE MIT RADIATION LABORATORY AND MICROWAVE RADAR

As 1941 dawned, the British were becoming increasingly desperate. German aircraft were bombing London nearly every night. English military forces struggled to mount an effective defense against the aerial invasion. After thousands of attempts, Britain's Royal Air Force had managed to hit fewer than ten enemy aircraft. It simply lacked the accuracy needed for nighttime fighting.

Meanwhile, in the United States, researchers at MIT's Radiation Laboratory, established just a few months prior, had been given the considerable task of devising an air defense system based on microwave radar technology. The system needed to work in real-world settings and at large scale. The scientists at MIT promised to have a test model of a new radar system in a bomber aircraft by the beginning of February 1941. But February arrived with little progress. On Wednesday, February 5, experimental physicist Luis Alvarez placed a bet that the device his team was developing on a rooftop overlooking the Charles River in Cambridge would soon be able to detect a passing airplane. Nothing worked that day or the next. Very early Friday morning, Alvarez and a few colleagues were back on the roof. One man hunted for airplanes through a telescopic sight while another pointed an antenna by hand and Alvarez stared at the radar scope. "Suddenly, a blip appeared," author Robert Buderi writes in his description of the event. "A bit incredulously, the two looked out from the penthouse to see a civilian plane flying nearby."[15]

Alvarez telephoned Radiation Lab director Lee DuBridge, reaching him during what had been a gloomy committee meeting. "We've done it, boys," DuBridge announced to the group. Less than two months later, a modified version of the new radar equipment, mounted in a B-18 bomber flying low across Long Island Sound, detected three American submarines several miles away. It was the first time anyone had detected a submarine with an airborne system.

The MIT Radiation Laboratory ("Rad Lab") eventually designed roughly 150 different radar systems, nearly half of all the radar used in World War II. Microwave radar systems were installed on planes and ships, in fixed ground installations, and on mobile trailers.[16] Systems developed at MIT helped Allied forces locate German submarines

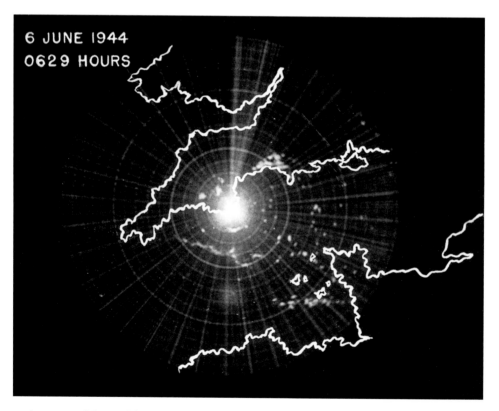

Radar image of the English Channel and the coast of Normandy, taken on June 6, 1944 (D-Day), the date of the Allied invasion of German-occupied France in World War II. *Source:* Courtesy, MIT Museum.

patrolling the Atlantic coast. They gave early warning of incoming enemy aircraft along the southern coast of England. They directed antiaircraft guns, allowed pilots to make blind landings on darkened or fog-shrouded runways, and guided fighters and bombers to distant targets. More than $1.7 billion worth of equipment (in 1945 dollars) was manufactured to designs originating in the Rad Lab.[17] Most of the lab's creations were assembled by outside contractors, but in some cases, the Rad Lab itself produced working units for immediate use by the military.

Origins of the Rad Lab

The MIT Rad Lab was launched in October 1940, shortly after a contingent of British military scientists led by chemist Henry Tizard traveled in secret to the United States to meet with American officials. Their aim was to discuss ways that the United States,

Worker assembling cavity magnetrons from stamped metal parts at the MIT Radiation Lab during World War II. *Source:* Courtesy, MIT Museum.

which had not yet officially declared war, could help its British allies. British physicist Edward Bowen carried with him precious cargo—a device called a cavity magnetron. An odd-looking grooved metal disc with pipes and wires protruding from it, the magnetron could generate radio waves at wavelengths of 10 centimeters or less, shorter than conventional radio signals and potentially useful in detecting enemy aircraft.

British and American scientists had been experimenting with radar (short for "RAdio Detection And Ranging") since the mid-1930s. The concept of radar is reasonably straightforward: bouncing radio waves off objects and detecting the reflected signal. The magnetron had recently been developed at the University of Birmingham, in England, where for the first time microwave energy was generated at sufficient power to be practical in radar systems.[18] But the British needed American engineering and manufacturing help to adapt it to a variety of sea, air, and land applications and to produce it in large enough quantities to be useful in the war.

"The magnetron, still an experimental device, had to be geared up for mass production," Buderi writes in a 1996 book.[19] "It needed to be combined with receivers, antennas, and a myriad of other components, and tailored to meet the various needs of the services—with different systems for airborne radars, submarine hunters, anti-aircraft guns, and bombers."

Though the details of how the magnetron actually generated radio waves remained a mystery even to scientists, it was incredibly powerful.[20] Luis Alvarez once said it was three thousand times better than previous devices. "If automobiles had been similarly improved, modern cars would cost about a dollar and go a thousand miles on a gallon of gas," he said in the mid-1980s. "Suddenly it was clear that microwave radar was there for the asking."

The Tizard delegation met with members of the National Defense Research Committee (NDRC), the advisory group established by President Roosevelt to coordinate scientific research and support America's defense. The NDRC agreed to launch a centralized effort to develop microwave radar in the United States. Some on the committee wanted the research to take place at the Carnegie Institution for Science in Washington.[21] But Vannevar Bush, head of the Carnegie Institution as well as the NDRC, worried about placing it at his own organization. "I protested and we had a hell of an argument that took half the night and a bottle of Scotch," he reportedly said. MIT President Karl Compton, a member of the committee, volunteered to host the effort in Cambridge. The group agreed on MIT because of its history in microwave research and because opening a new lab would attract little attention at a large research university. The MIT Radiation Laboratory—an intentionally deceptive name meant to suggest that the lab was exploring nuclear physics rather than microwave radar technology—officially opened a few months later. It was to have a profound effect on the course of World War II.

When it launched, the Radiation Lab focused on three areas of research: microwave-based aircraft detection, precision guidance for antiaircraft guns, and aircraft navigation.[22] Scientists in the lab often went out drinking on Friday nights, calling their social gatherings Project Four.[23]

Originally envisioned to have a staff of fifty, at its height in 1945 the lab employed nearly four thousand scientists, engineers, technicians, mechanics, stenographers, and other assistants.[24] The Rad Lab spread across the MIT campus, which has the quirky tradition of referring to its buildings by number rather than by name. The lab initially occupied portions of Buildings 3, 4, and 6. In August 1941, the Institute broke ground on another 50,000 square foot structure—Building 24—to accommodate the lab's growth. MIT soon added two more—Buildings 20 and 22—that together provided another quarter-million square feet of workspace. MIT operated temporary field stations across New England to test its air-, sea-, and ground-based radar systems. The Rad Lab became one of the largest civilian research efforts during World War II, second only to the Manhattan Project.

"A Legendary Rivalry"

In 1942, in a back corner of the Harvard campus, more than eight hundred scientists and support staff began working to develop radar-jamming technology, anticipating that Germany would develop its own systems.[25] The secret Radio Research Laboratory occupied a portion of Harvard's Biological Labs and two hastily built wood-frame buildings next to it.[26] The lab was led by Frederick Terman, who after the war was instrumental in creating Silicon Valley. Countermeasures developed at the Harvard lab included strips of metal foil that, when dumped from a plane, could mimic the radar profiles of additional aircraft, confusing radar signals. According to a 1945 article in the *Harvard Crimson*, the remnants were still visible at the war's end. "When American troops were occupying Germany after the surrender they were somewhat amazed at their discovery almost everywhere they looked of small glittering objects—narrow strips of bright metal foil which lays on the fields, in the roads, on the rooftops, and hung from the trees like the familiar tinsel on Christmas trees."

All told, Harvard's Radio Research Lab developed 150 different jamming systems. According to the Harvard Collection of Historical Scientific Instruments, which

Building 20 on the MIT campus, home of the Radiation Laboratory. *Source:* Courtesy, MIT Museum.

displays some of the technology developed there during World War II, "A legendary rivalry between the two laboratories was maintained through the war, as researchers at Harvard pointed their jammers across Cambridge towards MIT."[27]

Enduring Legacy

Microwave radar proved to be one of the most important technological advances of World War II. It was both the most widely used new technology and arguably the most effective in turning the tide of war. In November 1942, 117 Allied ships were bombed by German U-boats.[28] Less than a year later, after the Allies had installed many of the microwave radar systems developed in Cambridge, just nine of their ships were sunk in a two-month period.

The MIT Radiation Lab also developed a system of navigation for ocean-going vessels known as LORAN (long-range navigation), initially to aid naval convoys crossing

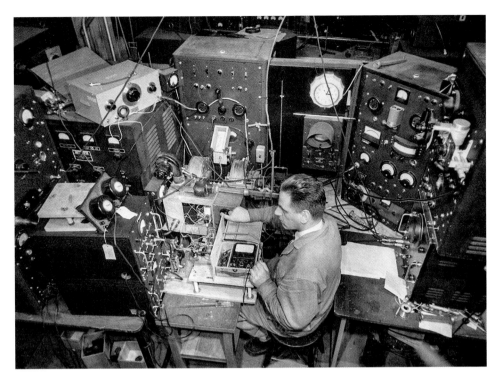

Testing equipment at the MIT Radiation Lab. *Source:* Courtesy, MIT Museum.

the Atlantic. A compendium of the Rad Lab's research issued in 1948 took up twenty-eight volumes. The lab's impacts on the American scientific community, on technological innovation, and on MIT and the city of Cambridge proved greater than just its role in the Allied victory in World War II.

The Rad Lab drew many of its thousand or so scientists, particularly physicists, from universities across the United States; MIT faculty members comprised only a small component of the staff.[29] Lee DuBridge, the lab's founding director, hailed from the University of Rochester. After the close of war in September 1945, most of these scientists returned home. Some of the most talented physicists in America worked at the lab during the war. At least eight members of the Rad Lab went on to receive Nobel Prizes in physics or chemistry.[30]

If there were doubts prior to World War II that government investment in academic research could produce tangible results, the work of the Rad Lab and the Manhattan Project laid those concerns to rest. The shape of the military-industrial-scientific complex, developed in an ad hoc manner to meet wartime needs, was firmly defined by the time the Rad Lab closed its doors at the end of December 1945.

MIT's defense funding—which had topped $93 million during the war[31]—was cut off immediately when the war ended but resumed a few years later with the onset of the Cold War against the Soviet Union. Close relations between the US government, university researchers, and defense contractors grew over the course of the 1950s and 1960s until these bonds were tested by popular disenchantment and unrest during the Vietnam War era.

One of the most direct commercial applications of the Radiation Lab's research was the microwave oven. Percy Spencer, an employee at the Raytheon Manufacturing Company in nearby Waltham, Massachusetts, is generally credited as its inventor. During World War II, Raytheon (founded by Vannevar Bush while at MIT) won a contract to manufacture radar sets designed at the Radiation Lab. Working on that contract, Spencer developed a faster method of making parts for the cavity magnetrons. At first, Raytheon was able to manufacture only seventeen magnetrons a day. Spencer figured out how to stamp them from sheets of metal rather than machining holes in solid blocks of copper, eventually increasing Raytheon's magnetron output to 2,600 per day.

Spencer happened to notice that the magnetrons heated up nearby objects (reportedly melting a chocolate bar in his pocket one day when he stood too close),[32] and he experimented with different foods until he perfected a cooking device.[33] Raytheon filed a patent application in October 1945 for the microwave oven and initially marketed it to commercial kitchens as the Radarange. Cavity magnetrons are still found in microwave ovens used today.

HAROLD "DOC" EDGERTON AND FLASH PHOTOGRAPHY

On a dark night in August 1944, Harold "Doc" Edgerton faced the famous Bronze Age monument at Stonehenge, England, braced his pocket camera against a fencepost, and clicked open its shutter. At a prearranged time, an airplane flew 1,500 feet overhead and momentarily flashed a 50,000-watt strobe light.[34] The camera captured an eerie image of the ancient circle of stones. Set against a darkened background, the monoliths appear to glow in the mysterious light from above.

Edgerton's mission was not simply artistic. In World War II, airplanes were used for the first time to take surveillance photographs of enemy troop movements. But these aerial reconnaissance missions were extremely dangerous for the pilots and crew. Photographing military targets often required two planes, one flying above the other. The top plane shined a continuous beam of light onto the ground while the second flew

Harold Edgerton in his MIT laboratory. *Source:* Courtesy, MIT Museum.

lower, taking pictures. When the lower plane was illuminated by the upper one, it was easy to spot—and shoot at.[35]

Through the experiments he conducted at Stonehenge, Edgerton, a professor of electrical engineering at MIT, proved that high-energy strobes could make these missions less risky. He synchronized powerful but brief flashes of light from a low-flying plane with the shutter of a camera mounted in the same aircraft.[36] By the time soldiers on the ground could direct artillery toward a plane, it would have long since disappeared into the darkness. Edgerton's instruments worked so reliably they were used to photograph the beaches of Normandy in the days leading up to the Allies' 1945 D-day invasion of France.[37]

Over his long MIT career, Edgerton created many memorable images with his strobe photography, including a golf club swinging in motion, a drop of milk landing, and a bullet piercing a balloon. Though his carefully crafted and often startling photos

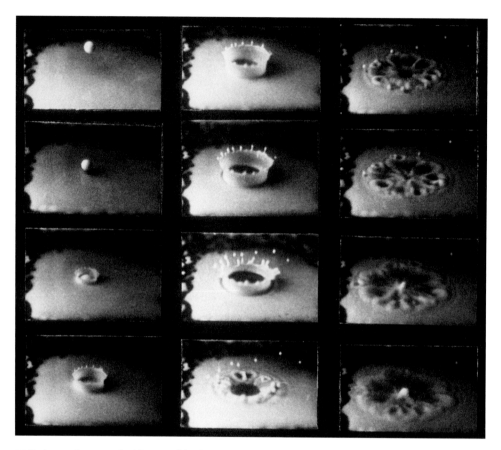

Milk drops photographed by Harold Edgerton with strobe equipment. *Source:* Courtesy, MIT Museum.

are worthy of art museums, Edgerton made them in the pursuit of decidedly practical ends. Descendants of his early strobe tubes still power office copy machines, mark the wings of airplanes, and pulse to the music of dance clubs. "Every camera today has a flash whose technological history—DNA—you could trace to Edgerton," says J. Kim Vandiver, a mechanical engineer, dean for undergraduate research at MIT, and director of the school's Edgerton Center.

Born in Fremont, Nebraska in 1903, by age twenty-eight Edgerton had earned his PhD from MIT and begun teaching there. Soon afterward, he produced the first completely electronic flash. The primary component was a sealed glass tube containing a gas (originally mercury vapor; later argon and xenon). When subjected to a high-voltage electric current, the lamp instantaneously produced bright light. As soon as the

Image of Stonehenge captured by Harold Edgerton while planes equipped with large strobe lights flew overhead, August 1944. *Source:* Courtesy, MIT Museum.

Stroboscopic photograph of balloons being pierced by a bullet, taken by Harold Edgerton. *Source:* Courtesy, MIT Museum.

current was removed, the light turned off, without the lag of conventional incandescent bulbs. The light discharge could be calibrated for the desired intensity and exposure time, and the device was rechargeable and reusable. Beginning in 1932, many of Edgerton's strobe devices were manufactured by the General Radio Corporation not far from Edgerton's MIT lab in Cambridge.[38]

Although he developed the first practical electronic strobe, Edgerton didn't invent the idea of flash photography. As early as the 1850s, pioneering photographer Fox Talbot experimented with artificial illumination, using magnesium-based powders to produce powerful sparks that lit up his subjects.[39] But the sparks were unreliable, loud, and uncontrollable. Other nineteenth-century photographers experimented with limelight, the illumination cast by heating calcium carbonate (lime) in a flame. This method often resulted in unflattering images. In the 1880s, German scientists developed Blitzlicht, a widely used flash powder composed of ground magnesium and potassium chlorate. An explosive mixture that could accidentally ignite, Blitzlicht claimed the lives or fingers of several unlucky experimental photographers. In 1929, the German Vacublitz was the first true flashbulb, made from aluminum foil sealed in an oxygenated tube and soon imitated by General Electric in the United States. Flashbulbs could be synchronized with the camera shutter to trigger simultaneously, and they were relatively cheap to produce. But they were single-use items; once ignited, they were discarded.

Edgerton initially used photography for industrial applications—to diagnose poorly performing motors, for instance, with carefully timed stop-motion images. His addition of synchronized rapid electronic flashes, also known as *strobes*, to high-speed photography made it more practical. With an abundance of charisma and passion, he became a tireless promoter of strobe photography. In 1933, he stuffed sixty pounds of equipment into the trunk of his car and drove halfway across the country—from Cambridge to his family home in Nebraska—stopping along the way to call on factory owners, demonstrating his technology and making sales.

Edgerton tried to convince Kodak, then the country's major photography equipment supplier, to manufacture his new strobe lights. According to Vandiver, Kodak's market research showed that the reusable flash would sell to only about fifty customers a year, and they declined to produce it. To generate publicity, Edgerton teamed up with an Associated Press sports photographer in December 1940 and captured heavyweight champion Joe Louis in the ring against boxer Al McCoy at Boston Garden.

In the 1940s and 1950s, Edgerton played a role in some of the largest flashes of light ever experienced on earth: the testing of atomic weapons. Edgerton, along with MIT colleagues Kenneth Germeshausen and Herbert Grier, invented trigger mechanisms used to spark nuclear explosions. The company they incorporated in 1947, EG&G, was involved in nine hundred tests of thermonuclear devices. They invented a camera that could photograph atomic blasts in carefully timed exposures lasting microseconds—millionths of a second.[40] In 1952, Edgerton travelled to the Marshall Islands in the Pacific Ocean to witness and photograph American atomic tests. "The electrical circuitry required to make a short-duration flash of light is essentially identical to the circuitry needed to detonate the bomb," Vandiver says.[41] Detonation required circuitry that could handle high current without burning up. Few others besides Edgerton, he says, could have designed such a detonator.

Edgerton combined science and art in his strobe photography, seeking not just a new tool to apply to industrial applications, national defense, and laboratory research but images that were pleasing to him. In the 1930s, he shot pictures of hummingbirds, frozen in action. He captured a bullet piercing an apple. No one had ever seen a drop of milk the way it appeared in Edgerton's photographs—a perfect white crown of liquid

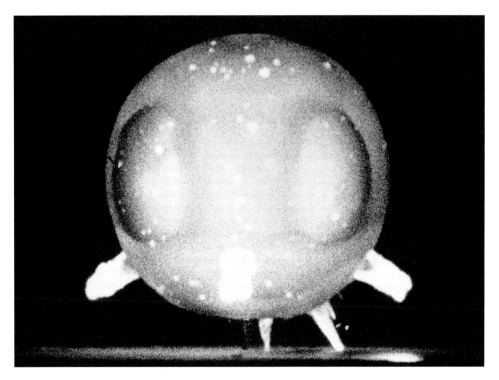

Image of nuclear detonation taken by Harold Edgerton, ca. 1946–1952. *Source:* Courtesy, MIT Museum.

frozen in time and set against a colored background. His most memorable images were hardly accidents; he worked with drops of milk for twenty-five years, Vandiver says, before creating his best-known shot in 1957. Edgerton "was a terrific photographer and had a super sense of composition," Vandiver says.[42]

Edgerton's photographs are in the permanent collections of a number of prominent museums. The Whitney Museum of American Art in New York City, which mounted an exhibit of Edgerton's work in 2018, noted: "Although uncomfortable being called an artist, Edgerton's work significantly expanded the legacy of such nineteenth-century figures as Eadweard Muybridge and Thomas Eakins, and shared some of the conceptual terrain of early 20th century movements such as Cubism and Futurism."[43] Vandiver remembers Edgerton once telling students, "I usually throw away all my bad negatives." Pointing to one example, Edgerton said, jokingly, "I almost threw this one away. But it's in the Museum of Modern Art now."

Throughout his life, Edgerton retained a childlike wonder and fascination with the physical world, and he was eager to share that enthusiasm with his MIT colleagues and students and anyone else willing to listen. "He was this gregarious, engaging, and charismatic guy," Vandiver says. If a student came into his office and wanted to try something, "Edgerton would basically say, 'There's a bench and a soldering iron. What are you waiting for?' He would essentially give his time, lab space, materials, and encourage students to try stuff out."[44]

Edgerton's lab at MIT became a must-visit destination for scientific researchers, artists, military brass—anyone who had a tricky problem to solve that might benefit from his strobe technology. The underwater explorer Jacques Cousteau became a friend and collaborator. Edgerton spent years developing an underwater imaging technology known as *side-scan sonar*, which is used by researchers, naval fleets, and shipwreck hunters to map the seafloor and locate lost treasures. Over his lifetime, Edgerton received forty-seven US patents for his inventions.[45]

Vandiver remembers first meeting Edgerton in 1972 when the strobe pioneer was nearing seventy. MIT had a mandatory retirement age, but Edgerton convinced the school to let him teach one class a year and maintain his lab without a salary. He taught a strobe project lab that decades later remains in the MIT course catalog.[46]

LOUIS FIESER AND NAPALM

For many Americans alive during the Vietnam War, the word *napalm* conjures up images of terrified villagers fleeing their burning homes, collateral victims of the US war effort in Southeast Asia. A photograph taken by Associated Press photographer Huynh Cong

"Nick" Ut on June 8, 1972, and published in Western newspapers the next day, became one of the enduring images of the Vietnam conflict. Ut's photo shows a nine-year-old Vietnamese girl—stripped naked as a result of an aerial napalm attack and screaming in pain as her flesh continued to burn—running down a road toward the photographer. The girl, Phan Thi Kim Phuc, would have died but for the extraordinary medical care she received at a clinic founded by Americans.[47] (She now lives in Canada and helps childhood victims of war around the world.[48]) The photograph, which earned Ut a Pulitzer Prize, helped turn napalm into an international pariah, associated in the public mind with gruesome burns, civilian casualties, and indiscriminate application. This was an outcome, it is fair to say, that was never anticipated by its inventor, Harvard professor Louis Fieser.

Fieser, an Ohio native, studied chemistry at Harvard as a graduate student, receiving his PhD in 1924.[49] After teaching at Bryn Mawr College in Pennsylvania for six years, he returned to Harvard as an assistant professor and spent the bulk of his professional career just off Oxford Street, retiring in 1968. Fieser was a productive research chemist, publishing 340 academic papers and thirteen books on organic chemistry, focusing on topics such as the synthesis of vitamin K, steroids, and antimalarial compounds that were especially useful to American soldiers serving in tropical areas during World War II.

In October 1940, as the United States readied itself to engage in the war then engulfing Europe, Fieser volunteered to conduct research on new military explosives under the aegis of the National Defense Research Committee (NDRC), led by former MIT dean Vannevar Bush.[50] In a basement laboratory in Harvard's Converse chemistry building, Fieser and his assistant E. B. Hershberg began work in June 1941 on a compound called divinylacetylene, testing its potential as a military explosive. The pair placed pans of the compound in their laboratory's window wells. They encouraged the chemical to thicken into a gel and attempted in vain to ignite it by throwing small stones at the pans and poking the gel with sticks. Lighting the gels with a match, the researchers noticed an "impressive sputter and sparkle." As Fieser recalled in a 1964 memoir, "We noticed also that when a viscous gel burns it does not become fluid but retains its viscous, sticky consistency. The experience suggested the idea of a bomb that would scatter large burning globs of sticky gel."

Fieser experimented with other chemical compounds, working under strict secrecy in his basement lab, often coming home covered in soot to his wife, who was unaware of his research.[51] During the summer of 1941, Fieser visited the Chemical Warfare Service at Edgewood Arsenal in Maryland, where he discovered that there was very little active research on incendiary bombs—munitions that would spread fire rather

Louis Fieser with laboratory equipment. *Source:* Harvard University Archives.

Louis Fieser's 1944 photo identification card from the National Defense Research Committee, certifying that he is authorized to consult with federal officials on the subjects of incendiaries and chemical warfare munitions. *Source:* Louis Fieser, *The Scientific Method: A Personal Account of Unusual Projects in War and in Peace* (New York: Reinhold Publishing, 1964), iv.

than simply explode on detonation. Returning to Cambridge, Fieser sought permission and funding from the NDRC to continue research on incendiary weapons, a request that was quickly approved. Harvard was awarded a contract worth $359,125 for "Anonymous Research No. 4." To manage the secret research effort, Fieser gave up his teaching duties and other laboratory projects, though he continued to draw a salary. Fieser's group moved to the third floor of Harvard's Wolcott Gibbs Laboratory, just off Divinity Avenue, where a special glass-enclosed room within a room was used to conduct tests of experimental incendiary materials. A hole was cut in the roof of the building, with a large fan that sucked smoke out of the test chamber after each burn. One of the compounds they tested was rubber cement, a common adhesive composed of natural rubber dissolved in a solvent. Commercial rubber cements contain nonflammable solvents; Fieser's group tested flammable solvents such as benzene and gasoline.

"Rubber-gasoline gels had the desired toughness and stickiness and rated nearly as high in incendiary effectiveness as gels in much less available solvents, and we settled on gasoline as the solvent of choice."[52] These compounds burned with intense energy.

Fieser's team loaded the rubber-gasoline gel into bomb cases and took them to an area near the Harvard Stadium, just over the Charles River in Boston, for firing.[53] The researchers stood atop the stadium to observe the tests, which they filmed for later presentation to their military sponsors. Once Fieser's group had developed a bomb with satisfactory performance characteristics, he brought a demonstration version to the Maryland arsenal, carrying the explosive device on a passenger train.

In late November 1941, days before the Japanese attack on Pearl Harbor, Fieser was asked by the US military for instructions on filling ten thousand bombs with his combustible gel, bombs that were then shipped to Manila in the Philippines.[54] The bombs were lost at sea, and American access to a supply of natural rubber was abruptly cut off by the Japanese.

The first field test of napalm, near Harvard Business School, July 4, 1942. *Source:* Louis Fieser, *The Scientific Method: A Personal Account of Unusual Projects in War and in Peace* (New York: Reinhold Publishing, 1964), 38.

Fieser immediately began the search for a substitute thickener. He turned to aluminum-based soaps, compounds known to form thick greaselike substances when dissolved in lubricating oils.[55] One promising formula that Fieser and his colleagues tested at their Harvard lab consisted of 5 percent aluminum naphthenate (a commercially available compound recommended by the Arthur D. Little Company of Cambridge), 5 percent of a commercial powder made from coconut oil known as aluminum palmitate, and 1 percent wood flour. The rest was gasoline. Combining names of two of the chemicals, naphthenate and palmitate, Fieser called the mixture napalm.[56] Though the chemical makeup of Fieser's incendiary gel was to change over time, the name *napalm*, like the material itself, proved sticky. He reported results to the NDRC in February 1942 and a decade later received US Patent 2,606,107.[57] The patent document notes that "the invention is concerned with the production of new and improved gelled hydrocarbon fuels and gelling agents therefor, for use . . . in flame throwers, in hand grenades, in fire starters and, generally, in any incendiary munition. . . ."[58] Continued testing in 1942 resulted in the creation of a brown powder that, when mixed with gasoline in the field, resulted in a material with the desired consistency and incendiary qualities.[59] Throughout World War II, Fieser continued research in his Cambridge laboratory on novel uses for his jellied gasoline, writing for the Office of Strategic Services, predecessor to the Central Intelligence Agency, a book titled *Arson: An Instruction Manual*.[60]

Napalm was first deployed in combat in August 1943 in Sicily. Though it continued to be used in the European conflict, jellied gasoline proved most effective in the war with Japan, where wood construction was more common. On February 25, 1945, American B-29 bombers dropped more than 450 tons of munitions, including napalm, on Tokyo, burning a square mile of the city and destroying some 28,000 buildings.[61] Less than two weeks later, 325 B-29 aircraft returned to Tokyo, dropping incendiary bombs and destroying a fifteen-square mile area at the center of the city. An estimated 100,000 people were killed in the firestorm.[62] It took weeks to clear the city of the charred bodies of the dead. Repeated napalm attacks on Tokyo and some sixty other Japanese cities by American forces continued until Japan surrendered in August 1945; the firebombing resulted in hundreds of thousands of civilian deaths.[63]

Napalm had proved to be devastatingly effective. For the American military, it had also been a bargain: the "Anonymous Research Project No. 4" cost only $5.2 million, compared with the $27 billion spent on the Manhattan Project that created the first atomic weapons.[64]

Napalm became a mainstay of America's conventional weapons arsenal in the second half of the twentieth century, used in the Korean conflict in the 1950s and again in the

Vietnam War in the 1960s and 1970s. Organized protests targeted the industrial companies that manufactured jellied gasoline for the US Defense Department, including Dow Chemical, which for a time was the compound's sole American supplier.[65]

In 1972, with America still entangled in the conflict in Vietnam, Fieser wrote to President Richard M. Nixon lobbying against using the jellied gasoline he'd invented. "It seems to me desirable to try to promote an international agreement to outlaw further use of napalm or napalm-type munitions," he wrote. The *Harvard Crimson* interviewed Fieser about the letter: "'I got the brush-off from Nixon,'" Fieser said.[66] The *Crimson* article noted that Fieser didn't regret having invented napalm; it was the chemical's use against people rather than buildings that bothered him. "'When we were developing napalm,' he says, 'we never thought of any anti-personnel use. We were thinking in terms of wooden structures, factories,'" he told the student-run newspaper.

Thirty-six years later, the US Senate and a newly installed President Barack Obama finally acted on Fieser's request, endorsing an international treaty on the use of incendiary weapons in civilian areas.[67] On his first full day as president, January 21, 2009, Obama signed into law Protocol III, part of the Convention on Conventional Weapons that covered devices "primarily designed to set fire to objects or to cause burn injury to persons through the action of flame, heat, or combination thereof."[68] The Obama administration reserved the right to use incendiaries if they would cause fewer deaths than another weapon. Previous presidential administrations had resisted signing onto the convention, first proposed in 1980, arguing in part that incendiary devices were the most effective weapons against certain targets, particularly bioweapons facilities where the biological material could be safely destroyed only with fire.[69]

Yet those involved in the convention continue to worry about incendiary weapons endangering civilians.[70] In 2016, white phosphorus, which also has incendiary effects, was used in Syria. "Air-dropped napalm is no longer the sole weapon of concern," according to a 2017 article about these devices. "Multipurpose and ground-launched incendiary weapons, which fall within Protocol III's loopholes, have become fixtures of contemporary armed conflict."[71] Support is building, the article said, for reviewing and strengthening Protocol III, but as of late 2020, it has not been changed.

CHARLES STARK DRAPER: INNOVATOR IN AERONAUTICAL NAVIGATION

In the early decades of aviation, prior to World War II, flying a small plane into a large cloud could be life threatening. Lacking a radio to call for help or accurate instruments for guidance, a pilot encountering a blanket of clouds was literally flying blind. MIT aviation pioneer Charles Stark Draper recalled what it felt like: "In several instances, a combination of poor weather and poor judgment left me that 'hopeless feeling,' under conditions of substantially no visibility for seemingly very long periods of time. During these periods I was lucky to remain alive."[72]

These experiences were harrowing but also motivating. Draper realized that flights would be much safer if they could be guided solely by instruments in the cockpit, and he spent much of his life perfecting navigation technology, including systems for ballistic missiles, airplanes, submarines, and the spacecraft used to take the Apollo astronauts to the moon and—just as important—bring them safely back to earth.

Draper's navigation systems were based on the principle of inertia, the idea proposed by Isaac Newton that a body at rest tends to remain at rest, while a body in motion tends

Charles Stark Draper. *Source:* Courtesy, The Charles Stark Draper Laboratory, Cambridge.

to remain in a uniform trajectory unless acted on by an outside force. Accurately measuring the inertial force on an airplane in flight, Draper realized, would allow him to calculate the speed of travel and distance flown. To do this, he combined several existing technologies. Gyroscopes developed by Elmer Sperry in the 1920s helped maintain airplane stability with respect to the horizon line but by themselves could not provide accurate information on a plane's position above the earth or its speed.[73] Sea captains had long ago learned to chart their trajectory with a knot meter, a compass, and an accurate clock.[74] Measuring four times a day might be enough to guide a slow-moving ship, but the pilot of an airplane traveling hundreds of miles an hour needs new calculations every second or so. Using gyroscopes and accelerometers, Draper devised systems to calculate a flight's path by automatically determining both position and speed.

A Missouri native, Charles Stark Draper first arrived at MIT in 1922 after graduating from Stanford University. He spent more than a decade as an MIT student—getting a second bachelor's degree, then a master's, and finally a PhD, though he ran an MIT lab before receiving his terminal degree.

In 1933, Draper established a laboratory to study problems with vibration in airplanes and engines.[75] During World War II, his Instrumentation Lab focused on creating accurate gun sights for antiaircraft weapons mounted on navy ships. Gunners firing on enemy planes from a rolling deck at sea used Draper's technology to stay focused on their targets.

After the war ended, Draper directed his attention to aeronautical guidance. In 1947, he began to develop on-board navigation systems for the US Air Force. His first device used celestial bodies, including the sun, as reference points to steer airplanes. This early system, which weighed 4,000 pounds, was tested in 1949 aboard an air force B-29.[76] Though crude, it proved the feasibility of Draper's idea.

Draper's next goal was accuracy. He aimed to build a navigation system that would guide a plane to within one mile of its intended landing spot after a ten-hour flight, this time relying solely on inertial rather than celestial guidance.[77] In February 1953, Draper—known as Doc to his colleagues—flew aboard a borrowed B-29 from Bedford, Massachusetts, to Los Angeles, with his inertial autopilot guiding the plane. The next morning, according to test pilot Chip Collins, Draper reported to an academic conference what had happened:

> . . . We all went to UCLA where industry presenters were scheduled to discuss the possibility of total inertial flight. Doc got up there and said we had just done it, and here was the evidence, showing them the track chart. "The accuracy," he said, "wasn't very good. We were off a few miles." There were audible gasps in the audience. People just couldn't believe it had been done.[78]

The Apollo Guidance Computer

Computer programmer Hugh Blair-Smith joined Draper's Instrumentation Lab in 1959 after graduating from Harvard.[79] Blair-Smith's first assignment was to work on a team creating a miniature onboard computer system that would navigate an air force spacecraft to Mars and back. The spacecraft's mission was to take a single photograph on film, fly back to earth, and discharge a cartridge with a parachute that could be picked up midair.

The Mars project was soon cancelled, however. Instead, the Instrumentation Lab received a contract to design the computer-controlled inertial navigation system that would guide NASA's Apollo flights to the moon. The Apollo Guidance Computer developed in Draper's lab was one of the first computers based on integrated circuits (microchips). The computer served the Apollo missions from 1966 to the mid-1970s.[80] It included optical equipment that astronauts could use to periodically recalibrate the inertial system, like sailors at sea navigating via the stars. It was important to NASA that the astronauts themselves command the spacecraft; its journey to the moon could not be completely automated. Draper engineers designed the computer with an input keyboard and display device for the astronauts, a form of human-computer interface that is now commonplace but at the time was novel.[81] During space missions, Draper staff monitored the Apollo computer's performance from an industrial building at the edge of the Charles River in Cambridge.[82]

Margaret Hamilton served as the lead developer for Draper's Apollo flight software. One of the few women involved in the massive Apollo project and the only one on the organization chart for Draper's leadership, Hamilton coined the term *software engineer* to describe the work she pioneered.[83] Hamilton says she was oblivious to the gender imbalance at the time. "We were all just working on something that had schedules and deadlines," she says. "Nobody seemed to care if you were a man or a woman. They only cared if you could help them with their assignment or deadlines."

Blair-Smith says Hamilton created an ethos of calm among the programmers. "I give her credit for a kind of serenity," he says. Every programmer encounters problems and then tries to replicate and fix them. Sometimes, though, the programmer can't recreate the problem, despite hours or days of effort. Hamilton called those FLTs—or funny little things. "You spend as much effort on dealing with them as seems to be required, but you must always be prepared for 'Okay, the consequences are not desired but not too bad either, so we'll focus on fixing the stuff that has more serious consequences,'" Blair-Smith says. "She taught us to take her own rather serene attitude toward 'We'll do everything we can because that's all we can do.'"

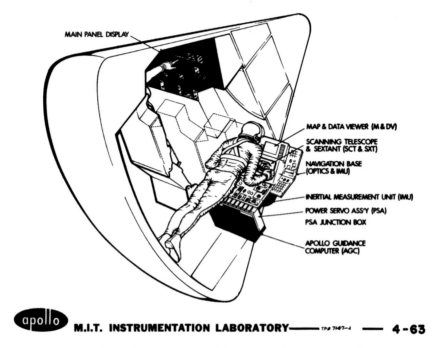

APOLLO G & N — SPACECRAFT LOCATION

MAIN PANEL DISPLAY

MAP & DATA VIEWER (M & DV)

SCANNING TELESCOPE
& SEXTANT (SCT & SXT)

NAVIGATION BASE
(OPTICS & IMU)

INERTIAL MEASUREMENT UNIT (IMU)

POWER SERVO ASS'Y (PSA)

PSA JUNCTION BOX

APOLLO GUIDANCE
COMPUTER (AGC)

apollo   M.I.T. INSTRUMENTATION LABORATORY——— TPø 7147-1 ——— 4-63

Diagram of the Apollo Guidance Computer's location within the spacecraft.
Source: Courtesy, The Charles Stark Draper Laboratory, Cambridge.

Hamilton occasionally brought her daughter to work, where the young girl loved playing with the Apollo flight simulator.[84] One day, she caused a crash. It took Hamilton a while to figure out what had gone wrong, but when she did, she brought her concerns to NASA administrators. "I went back to the management and said the astronaut could do the same thing. They said: 'It never will happen because he won't make any mistakes,'" Hamilton remembers. Sure enough, in 1968, Jim Lovell, when commanding *Apollo 8*, made the same mistake. She was soon allowed to fix the software to prevent it from ever happening again.

Other potentially disastrous mistakes—and tense moments—followed. During the aborted *Apollo 13* mission in 1970, most of the international media attention was focused on Mission Control at the Johnson Space Center in Houston, but Draper engineers and programmers in Cambridge played crucial roles as well.

On April 13, 1970, an oxygen tank exploded aboard *Apollo 13*'s Service Module, leaving the ship's other two components—the Control Module and the Lunar

Module—nearly crippled. The mission was quickly redirected from its initial goal—landing Americans on the moon's surface for the third time in history—to simply bringing the three astronauts back alive. Using only the rocket engine aboard the Lunar Module and with the advice of engineers on the ground, the astronauts had to maneuver their spacecraft into a trajectory they desperately hoped would lead them back to earth.

The key events of those days were captured in the 1995 movie *Apollo 13* starring Tom Hanks. Lance Drane, a Draper engineer who designed simulations for the Apollo missions, praised the film's accuracy, even though it ignored the role that Draper programmers played and the anxiety they suffered for four days until the astronauts splashed down safely in the South Pacific.[85]

"People on the ground were absolutely terrified. We were absolutely terrified," Drane says, comparing the improvised maneuvers with the Lunar Module to backing up an eighteen-wheeler while steering from behind. "The engine wasn't built for that kind of turn on and off," he says. Drane was part of the Draper team that ran computer simulations to prepare the crew for any possible eventuality—though not the one they eventually faced.

After Apollo

During the Vietnam War, antiwar protestors in Cambridge singled out Draper's lab for criticism because its funding and research interests were deeply connected to the needs of the US military. Student activists in 1969 noted that MIT was one of the Pentagon's largest contractors and that more than half of the Institute's annual budget was accounted for by just two labs, both involved in classified defense work: the Instrumentation Lab and the Lincoln Lab.[86] MIT students planned a protest and work stoppage on March 4, 1969, supported by some members of the faculty.

In 1973, the newly renamed Charles Stark Draper Laboratory became an independent research enterprise, separate from MIT. In 1976, Draper Lab moved into a new building nearby in Kendall Square. Draper engineers continued to develop missile guidance technology throughout the Cold War era. Among other projects, they also worked on the 1975 Apollo-Soyuz Test Project, in which American astronauts and Russian cosmonauts rendezvoused in space, and wrote computer code for the 1973 Skylab space station and the International Space Station, first launched a quarter century later. Demonstrations by antiwar activists continued at Draper's Cambridge headquarters into the 1980s.[87]

Draper himself remained involved with his namesake lab until his death in 1987 at age eighty-five.

Draper Laboratory, Cambridge, 2020. *Source:* Photo by the authors.

In addition to his prowess as an engineer, Draper was an inspiring leader. "He created a culture for people to work in that I regard as perfect," says Hugh Blair-Smith, "I've never had so much fun and excitement working anywhere else."

NOAM CHOMSKY AND THE RESPONSIBILITY OF INTELLECTUALS

The Bible was one of the top ten most cited sources in the arts and humanities between 1980 and 1992, according to Clarivate, a research data company.[88] Eight of the other slots were filled by people long dead and so familiar they are recognizable by just a single name: Marx, Lenin, Shakespeare, Aristotle, Plato, Freud, Hegel, and Cicero.[89] The tenth was Noam Chomsky.

Chomsky, a longtime member of the MIT faculty, gained recognition in the 1950s for his groundbreaking theories in the field of linguistics, the scientific study of languages. But it is his self-assigned role as a critic of American foreign policy that has brought Chomsky wide public notice. In the 1960s, he became a vocal opponent of US involvement in Vietnam, and he has continued in the decades since to lecture and write on what he sees as America's moral failings and ideological hypocrisy.

Undated photograph of Noam Chomsky. *Source:* Courtesy,
MIT Museum.

Although his own research was initially funded by the Pentagon, Chomsky has
labeled American military involvement in places like Vietnam, Iraq, and Afghanistan as
imperialist aggression. He has often called to attention what he sees as human society's
failure to address the existential threats of nuclear holocaust and global climate change.
Chomsky has maintained a role as a public intellectual and social critic despite a popular reception that has at times ranged from indifference to outright hostility. He has
lectured regularly over his long career, giving more than twenty interviews in 2017
alone[90]—when he was eighty-eight years old—and has written or contributed to more
than a hundred books, a third of them in the field of linguistics.[91]

"[Chomsky is] a chapter heading in the intellectual history of our time, whereas most of us are footnotes," the late Morris Halle, Chomsky's collaborator for six decades, used to say.[92] MIT, Chomsky's intellectual home from 1955 to 2017, was a good fit for him, says Samuel Jay Keyser, who headed MIT's linguistics department for two of those decades. The Institute cared more about the quality of his research than it did about his politics, making Chomsky an Institute Professor—the school's highest honor—at a time when he was highly critical of his colleagues in the political science department for supporting the Vietnam War. His breakthroughs in the field of linguistics, Keyser says, "made it possible for Noam to thrive here. MIT recognized that he'd done something special."

Noam Chomsky in a classroom at MIT, circa 1980. *Source:* Courtesy, MIT Museum.

Chomsky never joined a political party, and his views on the subject fall outside the traditional spectrum of American political thought. He has described himself as a "libertarian socialist," favoring a system that maximizes personal freedom and minimizes the power of the state but also opposes the capitalist private ownership of the means of production. To the extent that outsiders view Cambridge, Massachusetts, as "the People's Republic of Cambridge," a bastion of left-leaning activism, some credit may be due to Chomsky's sustained and vocal political dissent.

Origins of a Worldview

Avram Noam Chomsky was born in 1928 in Philadelphia, the son of Eastern European immigrants. At age ten, he wrote his first political treatise, a paper on fascism, following the fall of Barcelona to the forces of Francisco Franco.[93] He often said he developed his world view from the hours spent at his uncle's New York City newspaper stand as a teenager. "He encountered political ideas that were rampant in those years in New York City," Keyser says. "A lot of socialists, Russian Jews, Stalinists."

In 1951, after completing a master's degree at the University of Pennsylvania, Chomsky was admitted for a three-year term to the Society of Fellows at Harvard, a think tank within the university for young scholars of exceptional promise. After extending his stay at Harvard an extra year and completing a PhD at Penn, in 1955 Chomsky accepted a research position at MIT.

At MIT's Research Laboratory of Electronics, Chomsky operated outside the confines of an established university linguistics department, which suited him well.[94] Chomsky described MIT as "a pretty free and open place, open to experimentation and without rigid requirements. It was just perfect for someone of my idiosyncratic interests and work."[95]

Contributions to Linguistics

In 1957, Chomsky produced his seminal work, *Syntactic Structures*, which argues for the universality of grammar. In it, he promotes the idea that human beings are hard-wired to speak just as they are hard-wired to walk upright or see the world through binocular vision. This contradicted the prevailing theory of behaviorism, the idea that the human ability to speak is primarily a response to external stimuli. Babies were thought to be born as blank slates who acquired all of their language abilities in the first years of life. By contrast, Chomsky said that the rapid rate at which babies in every human culture acquire the capacity for language—in both linguistically rich and linguistically impoverished environments—suggests that language capacity is not learned but "grows" in the brain. Chomsky's ideas on natural language were foundational not only in theoretical linguistics but in the relatively young field of cognitive science, which studies the mind and its processes. "That idea of universal grammar didn't just change linguistics, it had repercussions for virtually every field that concerns the mind," New York University professor Gary Marcus wrote in a 2012 article.[96]

In an interview in his eleventh-floor Cambridge apartment, Samuel Keyser says that Chomsky's research suggested a new way of looking at the brain's evolution. He points out his window at the Boston skyline across the Charles River. "A gorilla is a magnificent creature, but it didn't produce what's out that window. Some change of circuitry in the brain did," Keyser says. "Chomsky, I think more than any person, focused on how important and yet how simple that change might have been."

Political Dissent

There is no sector of American society that has escaped Chomsky's sharp critique. Government leaders, whatever their political leanings or espoused ideology, represent in Chomsky's view the interests of powerful elites rather than the needs of the populace as a whole. "You don't trust government," he said in one speech.[97] "You don't trust authority. What you do is challenge authority." Corporations, he has argued, exploit labor and natural resources exclusively to the benefit of their shareholders, willfully blind to the longer-term human and environmental costs of their actions. Academics fail, according to Chomsky, when they don't speak out against what he sees as America's aggression abroad and its hypocritical justifications for international intervention. And he faults the mass media for amplifying elite perspectives and conveniently ignoring counternarratives.

Anti-interventionism, self-determination, the moral obligations of the powerful to the weak, and the role of commercial interests in American foreign policy have remained common themes in Chomsky's decades of political writing. To each of his essays and lectures he brings a trademark ironic humor and an abundance of supporting evidence—quotes and facts from news coverage of world events both current and historical. In the 1980s, he turned his attention to US involvement in the affairs of Central American nations, particularly Nicaragua, Guatemala, and El Salvador. In the 1990s, he criticized American action in Yugoslavia.[98] After the attacks of September 11, 2001, he opposed US military intervention in Iraq and Afghanistan.[99] He has received particularly severe criticism for his views on Israeli-Palestinian relations, where he has argued that Jewish treatment of Palestinians threatens to turn Jews into fascists.[100]

Noam Chomsky has written more than a hundred books since the 1960s. *Source:* Photo by the authors.

A *New York Times* columnist asked Chomsky in 2017 how he could stand to bear witness to so much human suffering.[101] "Witnessing it is enough to provide the motivation to go on," he replied. "And nothing is more inspiring to see how poor and suffering people, living under conditions incomparably worse than we endure, continue quietly and unpretentiously with courageous and committed struggle for justice and dignity."

In 2002, after twenty-six years as an Institute Professor at MIT, Chomsky retired with emeritus status. In 2017, he relocated to the University of Arizona, where he accepted a part-time professorship of linguistics and continued his engagement with the pressing sociopolitical issues of the day.

A large digital display screen dominates the intersection of Broadway, Main Street, and Third Street in Cambridge's Kendall Square, where Microsoft has its New England Research and Development Center. If you strain your eyes a bit, you can see Google's new location down Broadway, directly across from the newly built headquarters of Akamai, a content delivery company that hosts a significant fraction of worldwide internet traffic. In 2020, the area around Kendall Square housed one of the densest concentrations of computer application developers anywhere in the world. Amazon's Cambridge employees were working on Audible.com audiobooks and the Alexa virtual assistant; Apple engineers focused on Siri, speech recognition technology, and the field of machine learning; Google's coders worked on YouTube, the Chrome browser, flight search, and the Android mobile platform; IBM developed artificial intelligence products; and Microsoft's Cambridge teams advanced its Office 365 productivity software, mobile apps, and cloud services. Smaller tech companies were headquartered nearby, including HubSpot, a creator of marketing and sales software; Car Gurus, a comparison site for automobile shoppers; and Kayak.com, an online travel agent.

With MIT frequently ranking as the top undergraduate computer science program in the country by *U.S. News & World Report* and Harvard not far behind, it is hardly surprising that Cambridge became a global center of software development. But what may be unexpected are the sheer number of digital age milestones with Cambridge connections: several of the earliest digital computers ever operated (the Mark I at Harvard in 1944 and Whirlwind at MIT in 1951), one of the first video games (*Spacewar!*, developed at MIT in 1962), the first electronic message using the "**@**" symbol (sent in 1971 by a software engineer at Cambridge-based Bolt, Beranek, and Newman), the **RSA cryptography** system that allows secure financial transactions to take place over the internet, the first **electronic spreadsheet** programs (Visicalc and Lotus 1-2-3), and one of the first internet-enabled "sharing economy" businesses, **Zipcar**.

One of the earliest computing pioneers in Cambridge was Vannevar Bush, an MIT electrical engineering professor in the 1920s and 1930s, founder of the defense giant Raytheon, and science policy adviser to the federal government during the 1940s and

Postcard of the Massachusetts Institute of Technology, Cambridge. *Source:* The Tichnor Brothers Collection, Boston Public Library.

1950s. Bush invented a room-sized analog computer at MIT in 1927 called a differential analyzer. Perhaps just as important to the creation of the information age, while serving on the MIT faculty, Bush mentored both Frederick Terman, widely considered a cofounder of Silicon Valley, and mathematician/engineer **Claude Shannon**, whose 1938 MIT master's thesis introduced the fundamental concepts of digital circuit design.

Other computing pioneers with links to Cambridge include:

- Mathematician and theorist Norbert Wiener, who received his PhD at Harvard at the age of nineteen, in 1913. In a long career as an MIT professor, Wiener was influential in the fields of signal processing, robotics, automation, and feedback systems.
- Howard Aiken, a physicist at Harvard who partially designed and then managed the University's Mark I computer during World War II.
- Jay Forrester, who headed MIT's Cold War–era Whirlwind project and invented an early form of computer memory.
- Grace Hopper, a computer software pioneer who began her programming career at Harvard in the 1940s.

- Margaret Hamilton, who led the development of on-board flight software for NASA's Apollo program in the 1960s and who coined the term *software engineering* (see chapter 5).
- Robert Kahn, one of the inventors of the internet, who developed some of the network's key communication protocols in Cambridge in the late 1960s.
- Bill Gates, who attended Harvard for two years before leaving to found Microsoft in 1975.
- Sir Timothy Berners-Lee, who invented the World Wide Web's HTTP protocol and the first web browser in 1990 and has been affiliated with MIT since 1994.
- Rodney Brooks, an MIT robotics professor, a serial entrepreneur, and founder of iRobot, which developed the Zoomba robotic vacuum cleaner.
- Mark Zuckerberg, who started Facebook in his undergraduate dormitory at Harvard in 2004.

After World War II, MIT and Harvard spawned a host of computer hardware manufacturers, many of which settled in the Boston suburbs around Route 128, briefly dubbed "America's technology highway." Wang Laboratories, founded in Cambridge in 1951 by Harvard PhD An Wang, was an early maker of desktop calculators and word processing computers. In 1957, Digital Equipment Corporation, at one time the world's second largest computer maker, was started in Maynard, Massachusetts, by two MIT graduates. Thinking Machines Corporation, a supercomputer maker founded by an MIT graduate, was based in Cambridge from 1984 until its bankruptcy a decade later. Although Cambridge is no longer a leader in the computer hardware industry, it has played a key and ongoing role in the creation of the digital world.

CLAUDE SHANNON AND INFORMATION THEORY

We are surrounded by digital devices—computers, cell phones, televisions, household appliances, cameras, automatic teller machines. The list is seemingly endless. As recently as a century ago, however, the only common digital device was the light switch. A light switch has two states of being—on and off. No one imagined that light switches could carry data until an MIT graduate student named Claude Shannon published his master's thesis in 1938. Shannon argued that a set of simple mathematical equations called Boolean algebra could be modeled in electrical circuitry in a series of switches that are either in an on or off position. Shannon's observation was fundamental to the creation of the information age.

At the time of Shannon's breakthrough, the primary means of transmitting information long distance—the telephone and the radio—relied on what is called analog transmission, the manipulation of continuous electromagnetic waves. In a radio broad-cast, for instance, a microphone converts the sound of a person speaking or making music into continuously varied electrical signals, which are then amplified and trans-mitted over a band of the electromagnetic spectrum—that is, radio waves. After trav-eling great distances, these radio waves are converted back into sound energy at the receiving end.

Shannon ended up developing a completely different idea for transmitting informa-tion. He was influenced by the nineteenth-century English mathematician and logician George Boole, who adapted mathematical thinking to the field of logic. In Boole's math, every statement could be represented as either true or false. Shannon applied Boole's logical states to electrical circuits: true = 1 = on = a completed electrical circuit; false = 0 = off = a disrupted electrical circuit.

In 1936, Shannon moved to Cambridge from the University of Michigan, where he had studied electrical engineering and math. While he worked on a master's degree at MIT, Shannon joined the lab of professor Vannevar Bush.[1] Shannon's job was to operate Bush's differential analyzer, an early mechanical computer that could be used to solve complex math problems. The machine included an arrangement of gears, metal shafts, pulleys, and electrical switches and took up most of a room. It had to be completely reconfigured for each problem it was asked to solve.[2] While overseeing the differential analyzer, Shannon first saw parallels between Boole's idea of symbolic logic and the electrical circuitry that controlled the machine. He realized that electro-mechanical switches called relays, when arranged in complex circuits, could model the mathematical operations of the Boolean algebra he had learned as an undergraduate. Shannon's master's thesis, "A Symbolic Analysis of Relays and Switching Circuits," contained ideas that were startlingly new when published in 1938. Princeton mathe-matician H. H. Goldstine called the thesis "one of the most important master's theses ever written . . . a landmark in that it changed circuit design from an art to a science."[3]

Shannon explained how one might design an electric combination lock that would open only when its five pushbutton switches were pressed in the proper sequence.[4] Electrical circuits could be used to make decisions, such as whether to lock or unlock, based on complex inputs. This idea, first articulated by Shannon when he was twenty-two years old, is the fundamental concept behind all digital devices, whether they are made of vacuum tubes, transistors, or small squares of silicon. As early as 1886, Cambridge mathematician Charles Peirce (husband of Melusina Fay; see chapter 2) had described a connection between circuits and Boolean algebra.[5] Two key factors elevated the importance of Shannon's work: his timing, at a point when people could make technological use of his ideas, and Vannevar Bush's mentoring of the young man.

Claude Shannon. *Source:* Courtesy, MIT Museum.

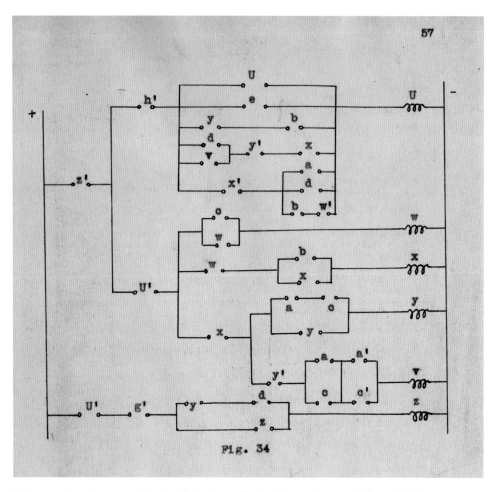

Diagram of an electric combination lock, from Claude Shannon's master's thesis, 1940.
Source: Courtesy, MIT Libraries.

Less than a decade after Shannon's paper, the first digital computers were invented. The Mark I, an elaborate electromechanical calculator, was built by IBM in Endicott, New York, and delivered to Harvard University in 1944; the ENIAC, the first general-purpose electronic computer, debuted in 1945 at the University of Pennsylvania; the Whirlwind I, the first computing machine with a magnetic-core memory, began operation at MIT in 1951.

In 1940, hoping to avoid military conscription as the United States prepared to enter World War II, Shannon went to work in New York City at Bell Labs, the research arm of American Telephone & Telegraph.[6] AT&T used profits from its near monopoly on US telephone service to underwrite basic research in physics, electronics, and other fields.[7] Shannon's math and engineering skills were well suited to some of the vexing technological problems America faced during World War II, including aiming antiaircraft guns at moving targets and transmitting information accurately to the battlefront without enemy interception.

In 1948, Shannon, still at Bell Labs, had another conceptual breakthrough. He published a paper laying out the fundamentals of what today is called information theory. Shannon insisted that any form of communication—a love letter, an audio transmission, a photograph, a movie—could be reduced to sequences of ones and zeros, the "on" switches and "off" switches of his MIT master's thesis a decade earlier. Shannon called these binary digits or bits. By converting information to digital bits, Shannon believed that it could be transmitted flawlessly, without the noise—like the static in radio broadcasts—that plagued analog transmissions. *Scientific American* magazine would later call Shannon's 1948 paper "the Magna Carta of the information age," establishing the fundamental principles that underlie modern computing.[8]

In 1956, Shannon accepted a visiting professorship at MIT and delivered a series of well-attended lectures. He had no curriculum for the class. He "simply invented something new for each lecture," says Robert Gallager, a former colleague of Shannon's at MIT.[9] "Most times when you go to hear a lecture by someone, at least if it's a technical lecture, after ten minutes, you're half confused and struggling to keep up. With Claude, it was never like that. You would listen to him give a talk: it was just simple, straightforward," Gallagher says. "Then he'd get this sly smile and out would come something absolutely fantastic that you would never have thought of, but you understood it."

Shannon's genius, Gallager says, was his ability to find the simplest way to look at everyday technological problems, whether communications, switching, cryptography systems, or solving mazes. He was not as much interested in practical applications as in addressing whatever issue was put in front of him. As he told an interviewer for an engineering journal in 1984:

> Bob, I think you impute a little more practical purpose to my thinking than actually exists. My mind wanders around, and I conceive of different things day and night. Like a science-fiction writer, I'm thinking "What if it were like this," or "Is there and interesting problem of this type" and I'm not caring whether someone is working on it or not. It's usually just that I like to solve a problem.[10]

Shannon returned to MIT as a full professor in 1958.[11] Exceptionally shy, he was uncomfortable in social situations and unwilling to promote his own achievements.

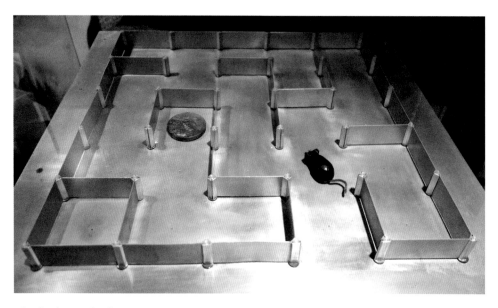

Claude Shannon's Theseus Maze, ca. 1952. *Source:* Wikimedia Commons.

Shannon formally advised only a few graduate students. He didn't like to write, Gallager says, and usually got his wife Betty, a mathematician in her own right, to take dictation and prepare his few publications.

During his two decades at MIT, Shannon gained a reputation for his eccentric hobbies: unicycling, juggling, and chess. "He loved to putter around, and he loved to build toys," Gallager says. Shannon famously developed a machine that consisted of a mechanical hand and a gearing mechanism. When the switch to the machine was turned on, the hand would move and turn off the switch.

Toward the end of his life, Shannon suffered from Alzheimer's disease. There were computers in his Winchester home, according to his wife, but "oddly enough, I don't think he even realized" the digital revolution he had spawned, she said. "That was the farthest thing from his mind—mostly he didn't pay attention to the outside world."[12] Shannon died in 2001.

INTERNET EMAIL

The way Ray Tomlinson told the story, he was essentially bored at work when he wrote a piece of computer software to send messages from one computer to another. This was in 1971, long before anyone used the words *internet* and *email*. Tomlinson, an engineer at the Cambridge-based consulting company known as Bolt, Beranek, and Newman (BBN), used an @ symbol to separate the name of the message's recipient from the host computer with which that person was associated. In the process, he is generally credited as having sent the first email message with the form of address we recognize today: name@host computer.

BBN was founded in Cambridge in 1948 by Richard Bolt and Leo Beranek, two MIT professors, and a third partner, Robert Newman, who joined in 1950. From their Harvard Square offices, they initially provided consulting services in architectural acoustics, a field that had been invented by Harvard physics professor Wallace Sabine a half century earlier (see chapter 4). Early projects included the United Nations' General Assembly building and concert halls at Lincoln Center in New York City.[13]

The company became a place where, in the decades after World War II, the cross-fertilization of different research disciplines—acoustics and psychology from Harvard, signal processing and radar from MIT—resulted in a fertile environment for new ideas.[14] Management oversight was intentionally loose, and entrepreneurship was encouraged. According to a company history, "The work was not directed in a top-down fashion; within broad limits, senior researchers were free to pursue their own research interests."[15]

By the late 1950s, seeing only a relatively small market for acoustical services, BBN expanded its research focus to computers. In 1957, Beranek hired MIT professor J. C. R. Licklider to lead its computing work.[16] A psychologist and theorist, Licklider promoted the development of time sharing, in which multiple people connect to the same computer, and he predicted a need for new human-computer interfaces that would be more intuitive to use and operate in real time.[17]

In 1962, BBN demonstrated an early time-sharing system, between computers in Cambridge and Washington, DC. A later version allowed doctors and nursing staff at the Massachusetts General Hospital to access patient information on a central computer from multiple locations.[18] As the authors of a company history note: "Users could now watch computers operate and begin to think about working with them cooperatively. Time-sharing was a major advance in human-computer integration and a sea change in the culture of computers and their users."

Licklider left BBN to join the Department of Defense's Advanced Research Projects Agency (ARPA), where, in 1963, he called for the interconnection of computer systems. Licklider saw computers not just as calculating devices but as communication tools. Licklider called this concept a "Galactic Network"—computers across the globe connected to allow anyone anywhere to quickly access information. He thus outlined the fundamental nature of today's internet.[19] The *New York Times* commented in 2012:

Ray Tomlinson. *Source:* Courtesy, Raytheon BBN Technologies.

"Though Dr. Licklider left in 1962, the company [BBN] became a favored destination for a new generation of software developers and was often referred to as the third university in Cambridge."[20]

The world of networked computing research consisted of a fairly small circle of people in the 1960s and early 1970s. In addition to Licklider, several of the field's pioneers had links to MIT or BBN, such as Lawrence Roberts, an MIT electrical engineering graduate often considered a "father of the internet."

Another was Robert Elliott Kahn, one of two people credited with the fundamental communications protocol of today's internet, TCP/IP (Transmission Control Protocol and Internet Protocol).[21] Kahn had been a young MIT electrical engineering professor when he moved to BBN in 1967. Reflecting on his time working at BBN, Kahn commented in 2016: "Leo [Beranek] and some of his colleagues had enough foresight to set up an institution that had many of the attributes of an academic institution. He created an environment in which people could explore ideas and take them much further than you typically did in an academic environment."[22]

After winning a research contract from the Defense Department in 1968, BBN began work on communication protocols to interconnect computers located in different locations, a network that became known as ARPANET (Advanced Research Projects Agency Network)—a forerunner to today's internet. BBN's task was to create a series of "message processors"—essentially network routers—that would serve as intermediaries between existing mainframe computers.

Tomlinson, hired at BBN a year before the ARPANET contract was inked, worked on software that could transfer files from one computer to another.[23] In 1971, when a researcher in California asked for advice on how to get messages to print on other computers, Tomlinson wrote software to accomplish something even better: sending messages directly to people rather than printers. His program combined two functions: rudimentary word processing and the transfer of files from one machine to another. The messaging system incorporated features that are now standard parts of an email header: the names of the sender and recipient, a subject line, and the date. To address his messages, Tomlinson searched the keyboard for a symbol to separate a user's name from a destination address. He settled on the @ symbol because it did not appear in anyone's name and did not already have a function in the operating system then being used on time-share computers.[24] In a 2009 interview, Tomlinson said of his program: "I worked on it for a few weeks, but it was not my main focus at the time. The overall software did not take much time at all, maybe three or four days total over a couple of weeks."[25]

A map of the ARPANET network showing the interconnections between computer systems at universities and military installations across the United States, March 1977. *Source:* Wikimedia Commons.

In late 1971, Tomlinson succeeded in sending a message from one computer to another. The two machines were only about ten feet apart in his Cambridge lab. "Because they were physically side by side, you could just sit at a table with two terminals on it and type something here and look over there and go back and forth. There wasn't a lot of running around to do. There were a lot of false starts. With any program, you write some code, you test it, it doesn't work. And if it works, you go home and celebrate because that's very unusual."

Tomlinson's messaging system initially linked only the thousand or so people in university and government laboratories who were connected to ARPANET, so it did not have a global impact for perhaps another decade. Nonetheless, it became popular within the ARPANET community. Tomlinson recalled:

[Electronic messaging] was adopted fairly quickly because it seemed to fill a niche that was needed. Telephone answering machines existed at the time, but they weren't common. If

The two machines used by Ray Tomlinson to send his first email via the ARPANET. In the foreground is BBN-TENEXA, the computer that received the first email. In the background is BBN-TENEXB, the computer that sent the first email. *Source:* Dan Murphy, Creative Commons Attribution-NoDerivatives 4.0 International License.

you were lucky, the person you wanted to contact had a secretary who would take a message or an answering service, and then they would call you back. But being able to communicate with somebody who was not ready to answer the telephone right then and there . . . there was nothing between that point in the spectrum and sending a letter by postal mail. E-mail became an instant hit for people who had to collaborate with others who were either, say, in a different time zone or just were not on the same schedule.[26]

Having developed many of the protocols of the government-run ARPANET, BBN created a commercial version called Telenet based on the same "packet-switching" communications protocols. Beginning nationwide operation in 1975, it was the first public computer network in the world based on these standards.[27]

Ray Tomlinson never set out to be a computer programming pioneer. A native of upstate New York, he attended Rensselaer Polytechnic Institute in Troy, completing a degree in electrical engineering before moving to MIT for graduate studies.[28] Originally planning to pursue a PhD in speech synthesis, Tomlinson became increasingly interested in computer programming. As he recalled later, "one afternoon I was walking down the hall in [MIT's] Building 26. There was this darkened room off the corridor with lots of noise coming out of it, yelling and screaming, and I looked in and saw about a dozen people all gathered around a CRT [cathode ray tube] playing *Spacewar!* And I said, 'This is really neat. I've got to learn how to program one of these things.'"

Tomlinson spent the remainder of his career at BBN. In 2012, he was inducted into the inaugural class of the Internet Hall of Fame, administered by the nonprofit Internet Society.[29] In its citation, the organization noted that "Tomlinson's email program brought about a complete revolution, fundamentally changing the way people communicate, including the way businesses, from huge corporations to tiny mom-and-pop shops, operate and the way millions of people shop, bank, and keep in touch with friends and family, whether they are across town or across oceans. Today, billions of email-enabled devices are in use every day." In 2010, the Museum of Modern Art in New York entered the @ symbol into its architecture and design collection, describing it as "a defining symbol of the computer age."[30]

PUBLIC KEY CRYPTOGRAPHY

Each time you buy something online or send money to a friend electronically, you're benefiting from public key cryptography, a system made practical in the 1970s by a trio of MIT professors to protect private communications from prying eyes.

The first computer networks, which were developed to connect researchers in government agencies and universities, were like public highways: anyone could see the data traveling along them. If you wanted to send confidential information electronically over the ARPANET or its successor, the internet, you needed a way to encrypt the data—to make it unreadable—and then to decrypt it at the end of its journey.

Encryption was used in both World Wars I and II to keep strategic information out of the hands of the enemy, but typically, code-breaking equipment had to be sent to the recipient ahead of the actual message. In the 1940s, Claude Shannon (discussed earlier in this chapter) was involved in cryptography efforts at Bell Labs, notably contributing to the development of an encrypted telephone hotline between President Franklin D. Roosevelt in Washington and Prime Minister Winston Churchill in London. Noting his simultaneous involvement in the field of information theory, Shannon said, "I started thinking about cryptography and secrecy systems. There is this close connection; they are very similar things, in one case trying to conceal information, and in the other case trying to transmit it."[31] The classified memorandum he wrote at Bell Labs in 1945, "A Mathematical Theory of Cryptography," was published in modified form a few years later. In his paper, Shannon proves that unbreakable cryptography is possible by using a random series of digits to encode the message so that the message itself is random. The random series of digits would be the "key" to unlocking the message.[32] The only issue was that both sender and recipient needed a copy of the key.

In the 1950s and 1960s, computer scientists began looking for ways to encrypt information sent between two strangers, such as a doctor and a patient or a retail business and a customer. In the early 1970s, a mathematician named Whitfield Diffie, working at Stanford University in California, began to focus on cryptography and its application in the networked computer age.[33] In 1976, Diffie and Stanford professor Martin Hellman published a paper called "New Directions in Cryptography" to propose the concept of "public keys"—digital certificates that could be sent to a recipient to safely unlock an encrypted message.

The concept works something like this. Suppose Alice in Massachusetts wants to send some very special chocolate chip cookies to Bob, who lives in California. (Cryptographers are fond of using examples with Alice and Bob). Alice could place a lock for which only she has the key onto a box containing the cookies and send them through the mail to Bob. The mail carrier can't open the locked box, but neither can Bob. So Bob places a lock of his own, to which only he has the key, onto the box of cookies, leaving Alice's lock on it as well, and he mails the box of cookies back to Alice. Now Alice can't unlock the box because Bob's lock is also on it, but she can remove her own lock, now that Bob's is securing the box. She then mails the locked box back to Bob, who then removes his own lock and eats some delicious (if slightly stale) chocolate chip cookies.[34] Even more simply, Bob could mail Alice an open padlock, to which only he knows the combination, and she could close the padlock on her box of cookies before mailing it to him. Instead of locks made of steel or brass, cryptographers use mathematical formulas that have special properties to scramble data they want to keep secret.

To encrypt information, Diffie and Hellman came up with the idea of multiplying two very large prime numbers (called factors) to obtain an even larger number (called the product).[35] In elementary school, most children are taught how to factor numbers: the number 6 has the factors 2 and 3, for instance, because 2 times 3 equals 6. But factoring, say, a 16-digit number like 2,345,678,987,654,321 would be difficult for most humans, and there are no known shortcuts to finding the answer. Multiplying prime numbers is a relatively straightforward process but going in reverse—finding the prime factors from the product—is considerably more difficult.

Turning Diffie and Hellman's abstract concept into lines of computer code was another thing altogether. To implement the Diffie-Hellman proposal, in 1976 three young faculty members at MIT—two computer scientists and a mathematician—began collaborating. Len Adleman, Ron Rivest, and Adi Shamir—all within a few years in age "and all passionate about the same topics," Adleman says—became great friends and collaborators.[36] The mathematician of the group, Adleman, describes this time at MIT as idyllic: "It was like living inside an early internet or Wikipedia. If you had a question, you didn't have to take more than three hundred steps, and the best person in the world to answer your question was there," he says.

Diffie and Hellman's paper lit a fire under Rivest and Shamir. As his friends became more interested in the problem, Adleman, who spent nearly every day with them, was dragged into their obsession, though he claims he didn't understand why they found it so fascinating. The three established a working method: Rivest and Shamir would come up with potentially unbreakable mathematical formulas, and Adleman would attempt to break them. This process went on for months.

One night in April 1977, the three attended a Passover seder in the Cambridge apartment of one of their graduate students. Adleman admits that his memory might be different than Rivest's, but he says that Rivest couldn't sleep after the seder and called him up sometime around midnight or early in the morning with an idea. "Ron

Leonard Adleman, Adi Shamir, and Ronald Rivest. *Source:* Ron Rivest.

says, 'Hey, Len, what about blah blah blah?'" Adleman remembers. "And upon hearing that, I said, 'Congratulations, Ron, I think you finally did it' because it looked solid to me. . . . I wouldn't know where to begin to break this."[37]

In February 1978, the three published "A Method for Obtaining Digital Signatures and Public-Key Cryptosystems" in an influential computer science publication, establishing their claim to the invention of practical electronic cryptography.[38] Adleman says he first tried to keep his name off the paper and then agreed to include his name last, rather than alphabetically, as Rivest had suggested. Rivest, Shamir, and Adleman described multiplying two 100-digit prime numbers, resulting in a 200-digit key; factoring this would prove a challenge even for a computer. From here, the math in their proposal gets more complicated but was still simple enough for the collaborators to describe in a four-page paper. In an early draft, they referred to their characters as A and B, but a reader suggested giving them real names: Adolph and Boris. Rivest instead chose Alice and Bob, and the names stuck.

The RSA algorithm, named for its inventors' last initials, was complex enough to secure data but also simple enough that it did not require lots of computer time to encrypt or decrypt a message. In the early 1980s, computers were not as powerful or as fast as they are now. Simplicity was critical to commercial viability. In 1983, MIT received a patent for their cryptographic system, and Rivest, Shamir, and Adleman formed a company—RSA Data Security—to market it. But who would find this innovation valuable enough to pay to use it? In a 2016 interview, Rivest recalled that there were initially few applications for the technology.[39] The company looked at ways to encrypt telephone calls, but "it turned out not to be where the market really was. That didn't work." In 1986, the company "just about went bust," Rivest said.

Luckily, by the late 1980s, computers were proving useful to people outside the scientific community. Everyday uses for computers, such as spreadsheets and word processing, were becoming common. The internet gained wider use because of email. Then, in 1989, Tim Berners-Lee, a British scientist working at European Council for Nuclear Research (CERN), invented what he called the World Wide Web.[40] Released for public use in 1993, the web made exchanging digital information much easier and led to new uses for RSA's encryption technology—namely, e-commerce. Amazon.com, eBay, iTunes, Zappos: none of these internet-based businesses could exist without secure digital transactions. Credit card numbers entered on e-commerce sites are still encrypted with variants of the technique that the RSA founders developed in 1977. Every website that starts with *https:* (as opposed to *http:*) uses public key cryptography. Even without considering the military applications of digital security, the commercial impact has been enormous.

United States Patent [19]

Rivest et al.

[11] **4,405,829**

[45] **Sep. 20, 1983**

[54] **CRYPTOGRAPHIC COMMUNICATIONS SYSTEM AND METHOD**

[75] Inventors: **Ronald L. Rivest**, Belmont; **Adi Shamir**, Cambridge; **Leonard M. Adleman**, Arlington, all of Mass.

[73] Assignee: **Massachusetts Institute of Technology**, Cambridge, Mass.

[21] Appl. No.: **860,586**

[22] Filed: **Dec. 14, 1977**

[51] Int. Cl.³ **H04K 1/00; H04I 9/04**
[52] U.S. Cl. **178/22.1; 178/22.11**
[58] Field of Search 178/22, 22.1, 22.11, 178/22.14, 22.15

[56]   **References Cited**

U.S. PATENT DOCUMENTS

3,657,476   4/1972   Aiken 178/22

OTHER PUBLICATIONS

"New Directions in Cryptography", Diffie et al., *IEEE Transactions on Information Theory*, vol. IT–22, No. 6, Nov. 1976, pp. 644–654.
"Theory of Numbers" Stewart, MacMillan Co., 1952, pp. 133–135.
"Diffie et al., Multi–User Cryptographic Techniques", AFIPS. Conference Proceedings, vol. 45, pp. 109–112, Jun. 8, 1976.

Primary Examiner—Sal Cangialosi
Attorney, Agent, or Firm—Arthur A. Smith, Jr.; Robert J. Horn, Jr.

[57]   **ABSTRACT**

A cryptographic communications system and method. The system includes a communications channel coupled to at least one terminal having an encoding device and to at least one terminal having a decoding device. A message-to-be-transferred is enciphered to ciphertext at the encoding terminal by first encoding the message as a number M in a predetermined set, and then raising that number to a first predetermined power (associated with the intended receiver) and finally computing the remainder, or residue, C, when the exponentiated number is divided by the product of two predetermined prime numbers (associated with the intended receiver). The residue C is the ciphertext. The ciphertext is deciphered to the original message at the decoding terminal in a similar manner by raising the ciphertext to a second predetermined power (associated with the intended receiver), and then computing the residue, M', when the exponentiated ciphertext is divided by the product of the two predetermined prime numbers associated with the intended receiver. The residue M' corresponds to the original encoded message M.

40 Claims, 7 Drawing Figures

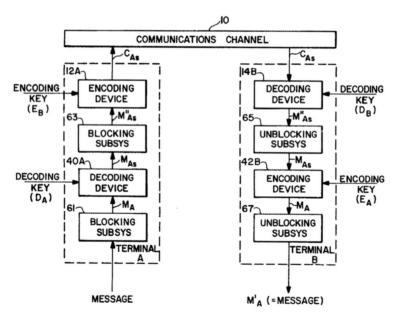

Cover of the US Patent issued on September 20, 1983, to Ronald Rivest, Adi Shamir, and Leonard Adleman for their Cryptographic Communications System and Method. *Source:* US Patent and Trademark Office.

At some future date, more powerful computers may be able to crack these algorithms, and newer encryption protocols will be needed. But Adleman doesn't think it'll happen any time soon. "The size of the number you could factor in, say, a day if machines went a million times faster might only be ten digits more than you could do on the old machine," he says. "Unless we find a really wonderful algorithm—and the current thinking is no such thing exists—factoring will always be hard."

THE ELECTRONIC SPREADSHEET

Anyone who has sat in front of a computer has probably used a spreadsheet program like Microsoft Excel or Google Sheets. Spreadsheets have become so indispensable—for keeping lists, managing budgets and business data, and making charts and graphs—that it's easy to forget that they did not always exist. MIT-trained computer scientist Dan Bricklin came up with the idea of the electronic spreadsheet in his first year of business school at Harvard in 1978,[41] and a few years later, Cambridge-based software entrepreneur Mitch Kapor ensured that the spreadsheet continued to be a "killer app" for newly introduced IBM personal computers, and for personal computing in general, for years to come.

In Bricklin's day at Harvard Business School, students would often fill large sliding blackboards with their calculations and models in analyzing a business problem. Sometimes, their math was wrong, or they wanted to change something, and it would take forever to fix by hand. Bricklin, sitting in one of his classes on the school's Boston campus, imagined a magic blackboard that would allow students to essentially do "word processing with numbers"—correct mistakes, account for changes, and make business decisions in real time.

Bricklin was bicycling on Martha's Vineyard during his 1978 summer break when he decided to turn his idea into a marketable product.[42] He needed a way to organize calculations in a way a computer could understand. He settled on the idea of a grid, where each numerical value would get its own box, allowing the computer to make calculations on the contents of that box.

Bricklin teamed up with a friend, Bob Frankston, and the two worked out a deal to commercialize the computer program. They engaged the help of another friend, Dan Fylstra, who ran a growing software company.[43] Fylstra and Frankston came up with the name of their new product—VisiCalc—while sitting at a restaurant on Massachusetts Avenue in Cambridge.

The next semester, in January 1979, Bricklin tested the software on a case study about Pepsi and presented it to his class. "I aced the case," he told an audience decades

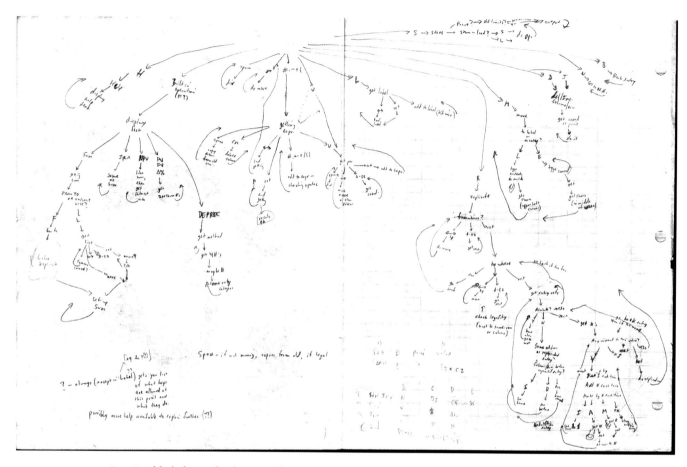

Dan Bricklin's design for the VisiCalc program, ca. 1978–1979. *Source:* Courtesy, Dan Bricklin.

later.[44] His professor wanted to know how he'd done so much work, but Bricklin wasn't ready to disclose his pet project.

With royalty payments from Fylstra's company, Bricklin and Frankston moved their young firm from Frankston's attic apartment in Arlington to Cambridge's Central Square.[45] In October 1979, they began selling their first commercial product, a version of VisiCalc designed to run on an Apple home computer. Apple founder Steve Jobs once told an interviewer that "VisiCalc propelled the success of Apple more than any other single event. If VisiCalc had been written for some other computer," Jobs continued, "you'd be interviewing somebody else right now."[46]

Large "mainframe" computers had been available commercially since the 1950s and were purchased by big companies and government agencies that could afford to devote

Vic's Egg on One on Massachusetts Avenue in Cambridge, where the first spreadsheet for home computers, VisiCalc, was named. *Source:* Courtesy, Dan Bricklin.

highly trained staffs and specially cooled rooms to operating the bus-sized machines. In the 1970s, the invention of the minicomputer brought the digital world within reach of midsized businesses and academic research groups. Then, in the late 1970s and early 1980s, the advent of the personal computer, most notably the Apple II and the IBM PC, changed how nonspecialists thought about computers and what they might be useful for. The personal-computer–based spreadsheet, both a powerful tool and easy to learn, accelerated that shift in thinking.

Mitch Kapor, an early personal computer software developer and industry insider, recalled that there wasn't anything else like VisiCalc and it had no precedent. "Dan and Bob deserve more credit than they get for having invented the spreadsheet. To me, it's right up there with the lightbulb. It should be a top ten invention of the century."[47]

Managers that Bricklin and Frankston hired got into a business dispute with a distributor, distracting them from product development, and although the pair created a version of their software for the new IBM PC, they were slow to take advantage of the machine's advantages over the Apple II, namely a more capable processor and larger memory.

Kapor sensed an opportunity. "I saw where the market was going," he says. In 1982, Kapor quit VisiCorp to set up his own company to create a spreadsheet program for the IBM PC. Kapor co-founded Lotus Development Corp. with Jonathan Sachs, a talented computer programmer. Sachs wrote the computer code for their first product over the course of ten months. Called Lotus 1-2-3, it represented a significant improvement over VisiCalc because it included graphics and database management features in addition to an advanced spreadsheet.[48] Lotus 1-2-3 was immediately successful, generating $53 million in sales in its first year. His spreadsheet, Kapor admits, "was very innovative, but it certainly didn't invent the genre. It just took it up to a much, much higher level and backed it up with good marketing and sales and support, so we became dominant." Kapor says he felt bad about damaging his relationship with Bricklin and Frankston. He later made it up to them by buying out their company, and they remain on friendly terms, Kapor and Bricklin both say.

Kapor, who grew up in New York and attended Yale University, landed in Cambridge when his college girlfriend got a job at the Boston public television station WGBH. She wanted to live somewhere she could get home delivery of the *New York Times*. His first choice was the West Coast, but Cambridge was acceptable because of its bookstores, intellectual life, and "stuff going on," Kapor says. As a recent college graduate, he couldn't afford Cambridge rents, so he lived in Watertown but spent his free time in Harvard Square. He later bought a house near the square.

The Boston area had long since become a hub for high technology. Spawned by research conducted at MIT and Harvard,[49] many of the state's high-tech companies settled in the Boston suburbs along the Route 128 corridor in the 1960s and 1970s. Minicomputer and hardware companies Data General, Wang, Prime Computer, and Digital Equipment Corporation together employed thousands of workers in the region.

In launching Lotus, however, Kapor bucked the suburban trend, deciding to base the company in East Cambridge, settling in an old factory building. "I wanted a place that did not require you to have a car," Kapor says, "which would make it easier for many different kinds of people to work there." He didn't like the malls, tract housing, and "soulless office buildings" of the suburbs. "I think we were the first tech company of any size to be in East Cambridge," he says. "I thought it would send a signal. It would be *the* place where we could get young, single people to work. I didn't know we were setting a precedent, but we were."

Almost overnight, Lotus became the third-largest personal computer software company in the United States. By 1984, the company had seven hundred employees, and the next year, it moved into a new nine-story headquarters on the banks of the Charles River in Cambridge.[50] By 1995, a dozen years after Lotus settled in Cambridge, the

Publicity photograph of Mitch Kapor with a Lotus 1-2-3 software box, ca. 1982. *Source:* Courtesy of The Mitch Kapor Archive.

city was home to two hundred software companies, and the state of Massachusetts as a whole employed nearly a hundred thousand in the software industry, according to the *New York Times*.

Lotus quickly found itself in the crosshairs of an aggressive rival: Microsoft, which developed the basic operating system software, MS-DOS, that ran the first IBM personal computers and later cemented its dominance in operating systems for personal computers by creating Windows. In response to competition, Lotus diversified its product line, creating an "office suite" combining its pioneering spreadsheet program

INTRODUCING 1-2-3.™ IT'LL HAVE YOUR IBM/PC JUMPING THROUGH HOOPS.

Advertisement introducing Lotus 1-2-3, ca. 1983. *Source:* Courtesy of The Mitch Kapor Archive.

Welcome to
THE LOTUS DEVELOPMENT BUILDING

Our new corporate headquarters is ready for us to move in! This brochure introduces you to the basic building layout and answers some questions you may have about location, security, amenities and space.

Cover of a brochure welcoming employees to the new Lotus Development Building on Cambridge Parkway in Cambridge, ca. 1985. *Source:* Courtesy of The Mitch Kapor Archive.

with word processing and database software. Lotus's bundling approach—familiar to contemporary users of the Microsoft suite, which includes Word and Excel—was not a commercial success. In 1987, Microsoft released Excel for Windows, and within only a few years, Excel was outselling Lotus 1-2-3.[51]

In 1989, just as the company's viability seemed particularly threatened, Lotus introduced Notes, an email and collaboration program that became widely popular, particularly in large corporate settings. Notes allowed a document to be opened on multiple computers across a network in real time, fostering electronic collaboration on complex projects. For example, a sales representative and her regional manager could simultaneously review quarterly sales projections from two different locations. When Notes was introduced in 1989, use of the internet was not yet widespread; the first browser for the World Wide Web wasn't invented until the next year. Less than a decade later, Notes had more than twenty million users.

Kapor left Lotus in 1986, as Notes was being developed. With the benefit of hindsight, he says, he can see that he was much better suited to starting companies than running them. "It doesn't play to my skill set to be managing a big company. I didn't give myself a gracious exit path, which I wish I had," he adds. More than the products he created at Lotus, Kapor says he thinks the company's lasting legacy was its culture. Employees were well compensated and received Lotus stock as part of their pay.[52] In an oral history interview, Kapor outlined some of the corporate values that Lotus put into practice: managers' bonuses were tied in part to employee ratings, an employee-run Philanthropy Committee decided where charitable giving would be directed, and the company took seriously the need to hire people with diverse backgrounds. Lotus refused to sell its products in South Africa during the apartheid era, and it was the first company to sponsor an AIDS walk.[53] "I'm terribly proud of the culture that we built and the impact it had on people," Kapor says.

In 1995, after stumbles in software development, late product deliveries, and several failed merger agreements, Lotus was acquired in a takeover by IBM. Seeking to further expand the market for Lotus Notes, IBM paid more than $3.5 billion for the Cambridge company, promising it autonomy in product development and keeping its brand name. Lotus continued to market its spreadsheet, word processing, and office productivity software.

For many of Lotus's 5,500 employees at the time of the acquisition (more than 2,000 of whom worked in Cambridge), the shift from the entrepreneurial environment that Kapor had created at Lotus to IBM's more established corporate culture was a difficult transition. Some left. In 2001, Lotus was largely absorbed into IBM's own software development operations after a corporate restructuring, and over the course of the next decade, Lotus reduced its Cambridge workforce.[54] IBM stopped using the Lotus brand name in 2013 and sold the business unit several years later. A pioneering company in the emergence of personal computing is now just a footnote of history, but the idea that originally launched it—the electronic spreadsheet–remains ubiquitous.

ZIPCAR

An argument can be made that America's "sharing economy" began in a neighborhood cafe in Cambridge's Central Square, where Robin Chase and Antje Danielson, the founders of a company called Zipcar, met for coffee in September 1999.[55]

Brochure for From All Walks of Life, an AIDS awareness walk along the Charles River in Boston and Cambridge sponsored by the Lotus Development Corp., 1986.
Source: Courtesy of The Mitch Kapor Archive.

Chase and Danielson often waited together outside their children's Cambridge school or watched their children on the playground. Occasionally, they talked about their career aspirations or the challenges of daily life.[56] One day, Danielson invited Chase to a local cafe to discuss collaborating on a new venture: a car-sharing service like one she had seen in her native Germany. Chase recognized that the internet and wireless technology could be used to make car sharing a viable business. The two realized that Cambridge—with its urban density and young, innovation-embracing population—might be a good testing ground for a new transportation concept.[57]

In 1999, there weren't many examples of businesses using technology to help people share—rather than own—assets. Craigslist, a cousin to the sharing economy, started in 1995, but operated exclusively in San Francisco until 2000. Airbnb, which opened homes and apartments to short-term rentals, launched in 2008; Uber began in 2009, allowing any car owner to become a part-time taxi driver.

In 2000, Chase and Danielson cofounded Zipcar, with Chase as CEO and Danielson, not yet able to give up her day job, as vice president.[58] Zipcar's convenient internet-based service offered its customers—the company called them "members"—the ability to reserve cars for as little as an hour of use. Members might borrow a Zipcar for a shopping trip, a work meeting in the suburbs, or a child's healthcare appointment. The company started with a fleet of four Volkswagens—two Beetles, a Golf, and a Passat—parked near transit stations in Cambridge and Boston.[59]

Chase said she knew that people would use a service like Zipcar because she wanted it herself.[60] Her husband used the family's car to commute to his job in the suburbs, and she often wished for a second vehicle —not every day, just once in a while. Cambridge is dense enough to traverse on foot, by bike, or via public transit most of the time, but many Boston area destinations are more conveniently reached by car. In a place where the cost of owning, insuring, and parking a private vehicle can run to thousands of dollars a year, Chase and Danielson suspected there was a strong market for car sharing.

As Chase recalled in her 2015 book, their instincts were right, but the initial payoff came even faster than she ever imagined:

> Hardly a minute after the Zipcar website went live (but before the launch), the phone rang.
>   "Hello, Zipcar. This is Robin. How may I help you?"
>   "Hi, I'd like to rent a car."
>   "Are you kidding me? We just went live! This is incredible! Sure!"
>   And so Craig Kleffman became the first Zipcar member. He rented our cars by the hour to transport his drum set to gigs he played at, and rented them by the day to get himself to the out-of-town triathlons he participated in. For people like Craig, who live in cities and don't need a car to get to work, both car ownership and car rental mean getting more

Andala Coffee House near Central Square in Cambridge, 2020. Zipcar founders Robin Chase and Antje Danielson met here to talk about car sharing. *Source:* Photo by the authors.

Robin Chase on the day Zipcar received one of its first vehicles—
"Betsy," a Volkswagen Beetle, 2000. *Source:* Courtesy, Zipcar/Robin
Chase.

car than they actually want to use. People chose Zipcar because sharing was the financially
smarter choice—and we were cool, smart, fun, urban, convenient and reliable as well.[61]

Whether or not it was the very first example of the sharing economy, Zipcar proved
to be an influential one, says Arun Sundararajan, a professor at New York University's
Stern School of Business.[62] "It's always been a remarkable company to me," Sundararajan says. "It's pretty clear that Zipcar in many ways was an idea that was a little ahead
of its time." Every company he spoke to in the early days of the sharing economy cited
Zipcar as an inspiration, "as a pioneer in demonstrating a new tech-enabled model of
shared consumption," he says. Zipcar opened the door to transportation services like
Uber and Lyft that make use of excess capacity: cars that would otherwise sit idle in
driveways or parking spots all day.

One of the ways Zipcar innovated was by using the internet to give its customers access to an asset in the real world—something that may seem obvious today but
wasn't in 2000, Sundararajan says. Car-sharing services existed in Europe and Canada,
Chase recalls, but they were cumbersome and didn't take advantage of technology at a
time when nearly half of Americans were already using the internet.[63] Communauto,
a car-sharing program in Canada, kept a paper ledger on each of its cars and wrote
in each reservation by hand. To collect the car, the customer had to locate a small
safe in the garage or nearby tree, plug in a combination, and claim the car key. Chase

and Danielson decided early on to enable Zipcar members to reserve their cars online and without the cumbersome check-in process of a traditional rental car company. The company used cell phone technology in each car linked to credit-card–like radio-frequency identification (RFID) devices that allowed a Zipcar to unlock for a member with a reservation.

Zipcar was "paradigm shifting," says Glen Urban, a marketing professor at the MIT Sloan School of Management who advised Chase early on.[64] "There's big innovation here: [the automobile] is a product that's psychologically identified with ownership," Urban says. For many Americans, the car is a status symbol, "the fulfillment of a dream of success." For Zipcar to succeed, customers had to be willing to think of it less as a reflection of personal worth or style and more as a service, one that was less expensive than a long taxi ride, more convenient than a city bus, and easier to book than the car you might rent on vacation.

Chase sees similarities between what she and Danielson built with Zipcar and platforms like YouTube, Airbnb, and Waze, in which users share content with strangers to create something new of value.[65] "There is one structure that underlies all these" she says, "excess capacity + a platform for participation + diverse peers—and it is fundamentally changing the way we work, build businesses and shape economies."

Being at the cutting edge of technology wasn't easy for either Danielson or Chase, particularly in a business—the auto industry—long dominated by men. Both benefited financially from their ownership stakes in Zipcar, but neither spent long with the company. Danielson left in January 2001, joining the Tufts University Institute of the Environment and later, the MIT Energy Initiative.[66] Chase served as Zipcar's CEO through February 2003 and sat on the company's board of directors for two more years before departing. Since then, she has become a serial entrepreneur in the transportation field, frequent public speaker, and author.[67]

In an interview, Chase recalls one pivotal moment in Zipcar's launch. Three days before the car sharing service was set to open for business, the company that had agreed to lease Zipcar three of its first four cars demanded deposits of $7,000 for each car. Zipcar had just $68 in its bank account.[68] As it happened, Chase went out that night to the launch party of another startup. There, she ran into Juan Enriquez, an angel investor she had been trying to convince to support Zipcar. "How are things going? What can I do for you?" Enriquez reportedly asked her. Chase responded: "I need $25,000 by tomorrow morning." "Done," he replied. The money was in the company's bank account the next morning, and its web site went live a few days later.

There were other hiccups. A few months after Zipcar's launch, Chase faced another big challenge: she realized she had priced the daily rentals too low.[69] If she didn't do

A Zipcar on display at the Harvard Science Center plaza, 2016. *Source:* Photo by the authors.

something, the company would quickly go bankrupt. After a few hours of agonizing, Chase drafted an email to the company's "members," confessed her mistake, and explained why she was raising daily rates by 25 percent. By the next day, she had received a few dozen replies, only two of which were angry.

Danielson had her own trials by fire. One night in the company's early days, she had to respond to a Zipcar with a dead battery. She described going out at midnight to set the jumper cables while her newborn slept in a car seat.[70]

Chase says the pair benefited from the intellectual resources available in the Cambridge area. Urban, her former MIT Sloan professor, advised Chase to think beyond the Boston area and reach markets across the country. "I thought the idea was great," he says. He encouraged her to "jump for the market as opposed to slow evolution and roll-out." Chase also got advice from professors she cold-called, including several at Harvard Business School who helped her think through a pricing strategy for Zipcar.

Chase says Cambridge was the right place to start Zipcar because of its density and population. Because the city is relatively small, she knew intuitively where to place the first cars. "The scale of it is something you can get your hands around," Chase says, describing Cambridge, where she still lives, as the minimum viable city size for a service like Zipcar to take off. "Cambridge really has the attributes for experimentation of what it is to be a city, but it's not Brooklyn or Manhattan," she says. "It's a friendly and relevant size."

Zipcar did expand beyond its Boston-area roots and became a publicly traded company in 2011, raising $173 million on its first trading day.[71] Car rental giant Avis bought Zipcar for $491 million in 2013, the first year it posted a profit.[72] Danielson said she started off with 50 percent of the company but ended up with just 1.3 percent, earning about $6 million from the sale. Chase ended with a larger stake, though still modest as a percentage of the overall company.[73] As of 2020, Zipcar, headquartered in Boston, called itself "the world's leading car-sharing network" and operated in more than three hundred American cities and at least that many American universities, as well as six other countries.[74]

In 2001, Swiss pharmaceutical giant Novartis hired Mark Fishman to suggest ways it might rethink its approach to research and help scout possible locations for a new global research headquarters in the United States. In his conversations with CEO Daniel Vasella, Fishman, a cardiologist and researcher at Massachusetts General Hospital, recommended Cambridge. It made good business sense to put the company's research arm next to two of the world's greatest research universities and near some of its best hospitals. Plus, the presence of locally grown companies Biogen and Genzyme showed that there was strong scientific and business talent Novartis could tap. The time zone and travel distance to the company's European headquarters made the Boston area more convenient than the West Coast. Vasella liked Fishman's idea so much that he hired him to run Novartis's new research arm in Cambridge. The company renovated the former Necco candy factory on Massachusetts Avenue, a short walk to MIT and a quick bus ride to Harvard and Boston's major research hospitals. Pierre Azoulay, an economist at MIT who studies biomedical innovation, says Novartis—and later other pharma companies—realized that "one way to rejuvenate the discovery research piece [of their industry] was to locate in an ecosystem."[1]

Fishman's simple idea—that interactions with people from different companies and different areas of expertise would spur innovation—triggered the city of Cambridge's exceptional economic boom in the early twenty-first century. Within fifteen years of Novartis's arrival, thirteen of the world's top twenty pharmaceutical companies opened major offices or were headquartered in Cambridge.[2] Public funding followed, with the state investing $1 billion in the region's biotech industry in 2008 and extending that contribution a decade later.[3]

Novartis benefited as well. Between 2002 when Fishman started there and 2016 when he returned full-time to his own research at Harvard, the Novartis Institutes for BioMedical Research discovered and won approval for ninety new medicines designed to treat more than 120 conditions.[4]

Pharmaceutical researchers arrived in Cambridge's Kendall Square at a time of major advances in how scientists understand biology at the molecular level—advances that

Aerial view of Kendall Square, 2016. *Source:* Alex MacLean.

have fundamentally transformed medicine and drug development. "In the 20th century, we treated diseases based on their symptoms. What's happened in the 21st century is all of medicine is getting rethought in terms of the fundamental causes of disease, the genes that are mutated," according to Eric Lander, who left his position as head of the Broad Institute of Harvard and MIT, a genetics research center, in early 2021 to join President Joseph Biden's cabinet.[5] Science takes place in many parts of the world, but Lander and others in Cambridge have played an outsized role in that transformation—identifying the genes and explaining the cellular mechanisms that cause disease, as well as developing the therapeutics that can treat or even cure them.

Origins of a Microbiology Powerhouse

Cambridge wasn't always a center of medical innovation. Harvard Medical School got its start in Harvard Yard in 1782 with medical lectures given at Holden Chapel, but the school moved to Boston in 1805 so its professors, who were also practicing physicians, could be closer to the bigger city's hospitals and patient base. Boston earned renown as a center of innovation in health care for such early activities as the first use of ether as an anesthetic in 1846 and the development of infant formula in 1919.

Matthew Meselson at his offices in the Biological
Laboratories of Harvard University, 2019.
Source: Photo by the authors.

Harvard's biology department was largely focused on whole organisms in 1956 when James Watson arrived there, fresh from describing the structure of DNA.[6] Watson's vision of what a biology department should be was very different, and his brash style and insistence that biology's future lay at the molecular level did not win him a lot of friends. But he soon brought in colleagues like Mark Ptashne, Walter Gilbert, and **Matthew Meselson**, who all made huge contributions to the field's early days.

In the summer of 1960, at Cold Spring Harbor on Long Island, New York, newly minted college graduate David Baltimore met Salvador Luria, a pioneer in molecular biology. Baltimore was looking for a job, and Luria made an offer.[7] "He said I'm going to MIT. I'm going to start up molecular biology. Do you want to come?" Baltimore remembered. "The amazing thing is there wasn't anywhere else to go. There weren't departments of molecular biology in 1960."

Luria had been Watson's PhD thesis adviser at Indiana University and brought the same zeal for molecular biology when he joined his former student in Cambridge. An Italian Jew, Luria had escaped Vichy-controlled Paris on a bicycle during World War II and shared a Nobel Prize in 1969, helping to cement MIT's reputation in the field of biology. Three people he hired—Baltimore, Susumu Tonegawa, and Phillip Sharp—went on to win their own Nobels, and MIT became a leader in unraveling the basic biology of cancer, a role it continues to play today. Har Gobind Khorana joined the biology faculty in 1970, after doing groundbreaking work helping to explain how RNA codes for the synthesis of proteins. Over the next six years, Khorana and several colleagues were the first to synthesize a gene—marking the beginning of the ability to write genetic code as well as read it.

Ecosystem in Action

The Ragon Institute—a joint effort of Mass. General Hospital, MIT, and Harvard—exemplifies the type of academic collaborations that are now common in Cambridge. Ragon, which aims to harness the power of the immune system, was founded in 2009 with a $100 million donation from the owners of a Cambridge-based software company. They added another $200 million a decade later. The institute's scientific founder, Bruce Walker, of Harvard, made his reputation studying extremely rare HIV patients whose immune systems are able to control a virus that in others rages out of control. Under Walker, Ragon expanded to include other immune-related conditions. One affiliated researcher, Dan Barouch, conducted the basic science behind Johnson & Johnson's COVID-19 vaccine.[8]

The experience of MIT bioengineering professor Sangeeta Bhatia provides a different example of what she calls the "really magical" biotech ecosystem in Cambridge.[9] In 2015, she came up with the idea for a company and went to visit **Robert Langer**, an MIT professor and serial entrepreneur whose lab is two floors above hers at the Koch Institute for Integrative Cancer Research at MIT. A few days later, she and Langer were at the same social event, and Langer encouraged her to describe her idea to another guest, venture capitalist Terry McGuire, who has backed about two dozen of Langer's forty-three companies. "So I did. And he got excited," Bhatia said. The next day, she walked across the hall to Nobel Laureate Phil Sharp's office. "And I said: 'Is this a good idea, what do you think? Can you introduce me to somebody?' And he did."[10] One of her postdocs was willing to leave MIT to lead the company. The Institute's technology licensing office, which helped Bhatia obtain patents for her ideas, was a short walk away. "And there's the incubator space right up the street where you get the space," she said. "That kind of density—you start bumping into everything you need to make it happen in short order. I think it's a real accelerator."[11]

William Sahlman, a professor of entrepreneurship at Harvard Business School, says he's never seen a biotech ecosystem function better than Cambridge's.[12] "This whole thing is running about as well as any system you could name," he says. "Even better than Silicon Valley, in my view, right now." There aren't as many scientists in Silicon Valley focused on starting companies, Sahlman says, rattling off the names of a half dozen Cambridge-based entrepreneurial scientists and noting that the thousand researchers affiliated with the Harvard Stem Cell Institute outnumber all the stem cell scientists in the state of California. "We just have an engine that is moving science forward at an extraordinary rate," he says. "The consequences for human health will be remarkable."

In this chapter, we feature five facets of this environment, each of which both benefited from and added to the richness of Cambridge's biology ecosystem. First, we take a look at the career of Matt Meselson, who was already a giant in the field of molecular biology when he arrived at Harvard in 1960. As important as his contributions to the field of biology have been, though, has been his role in protecting the world from biological and chemical warfare.

In some ways, a 1976 dispute over **recombinant DNA and biosafety** undergirds all of the biotechnology and pharmaceutical development that has come to Cambridge since. Restrictions imposed on local research by a zealous city leadership paradoxically provided clear ground rules for undertaking biotech work here. In the earliest days of the biotechnology industry, Biogen cofounders Phil Sharp of MIT and Wally Gilbert of Harvard relocated the company to Cambridge because of that predictability and the sheer convenience to their research labs.

About a third of the **Human Genome Project** took place in Cambridge. Scientists here were among those who first conceived of the then crazy idea of mapping every human gene, and diligence and good luck put them ahead of many of their peers. The infrastructure initially put in place for that project now supports millions of dollars a year in genetics research. And recent advances in the **tools used for gene editing**—made at least partly in Cambridge—are transforming the practice of medicine and drug development.

Finally, at MIT, Robert Langer has almost singlehandedly spawned entire new fields and industries and exemplifies the interconnections among the city's academic, financial, and government sectors. Langer's combination of interests in biology and engineering, which colleagues early in his career derided as a waste of time, has led to advances in drug delivery, tissue engineering, material science, and vaccine technology.

Engineers had long made medical machines, like heart pumps, but Langer was among the first to try to engineer biology itself—to change the molecules that could penetrate a cell or grow replacement parts in a dish. "He's the father of biomedical engineering in many ways," says Ali Khademhosseini, a bioengineer who completed postdoctoral studies in Langer's lab.[13] "He was becoming interested in medicine before any other real engineers were into that."

MATTHEW MESELSON: MAKING THE WORLD A SAFER PLACE

Well into his late eighties, Matthew Meselson still has a spacious corner office and a research lab in the Biological Laboratories at Harvard. An unassuming man with a quick wit and a prodigious memory, Meselson lines his office walls with mementos from decades of work and three-ring binders with laser-printed titles such as "Yellow Rain Vol 2." In 1957, Meselson and Franklin Stahl were the first to prove experimentally that James Watson and Francis Crick were correct in describing the shape of DNA, the carrier of genetic information, as a double helix. A few years later, Meselson helped discover messenger RNA, a key step in the process of turning genetic blueprints into actions. In the late 1960s, his work led to new tools for cutting and splicing genetic material. And seemingly in his spare time, Meselson has helped protect the world from biological and chemical weapons.

At a celebration of Meselson's eighty-eighth birthday in 2018, former US Secretary of Defense Ashton Carter noted that "Matt, more than any other person in the world, is responsible for the Biological Weapons Convention, which expresses the abhorrence of humankind towards the weaponization of biology, of life itself being used against life."[14]

Meselson got his first big break as a scientist in 1953, the summer before heading to graduate school. He was at a swimming party at the house of his California Institute of Technology classmate Linda Pauling, daughter of biochemist Linus Pauling. As Meselson remembered it, when he was in the pool, "the world's greatest chemist comes out wearing a tie and a vest and looks down at me in the water like some kind of insect."[15] Pauling asked what the young man was doing for graduate school. When Meselson mentioned the name of the professor he planned to study with in Chicago, "Linus looked down at me and said, 'But Matt, that's a lot of baloney. Why don't you come be my graduate student?' So, I looked up and said 'Okay.' That's how I got into graduate school."

In 1954, Meselson spent the summer at the Marine Biological Laboratory in Massachusetts, working as a lab assistant for James Watson, who had described the structure

President Gerald R. Ford at the East Room signing ceremony for the US instruments of ratification of the Geneva Protocol of 1925 and the Biological Weapons Convention, January 22, 1975.
Source: Courtesy Gerald R. Ford Presidential Library.

of DNA with Francis Crick only the year before.[16] It was there that Meselson met another graduate student, Frank Stahl.

Meselson told Stahl he was hoping to figure out if Watson and Crick's theory of DNA replication—then a controversial new idea—was correct. Meselson had an idea for using differences in densities to test models for DNA replication. In 1957, after several years of collaborating, Meselson and Stahl finalized their experiment, which was published the following year. "They designed an experiment so elegant and so decisive it became known as the most beautiful experiment in biology," Thomas Maniatis, chair of the department of biochemistry and molecular biophysics at Columbia University, said in presenting Meselson with the prestigious Lasker Award in 2004.[17] Their work showed that as DNA replicated, it produced two identical "daughters," which each contained one parent and one new strand of DNA. As Meselson noted in a 2003 interview, "for some people, [the experiment] made the double helix seem more concrete, less abstract."[18]

In 1960 while still at CalTech, Meselson worked with other researchers to demonstrate the existence of so-called messenger RNA (shortened as mRNA), which helps create proteins from a DNA blueprint. This research supported a theory of Francis Crick's that became known as the "central dogma" of genetics: DNA makes RNA, which in turn makes the proteins that perform all the activities of life.[19] The results of the research by Meselson and his collaborators were both novel and influential when they were published in *Nature* in 1961, alongside a paper from Harvard's Jim Watson and Wally Gilbert that had a similar finding.[20]

That year, Meselson moved from California to the biology department at Harvard, which would become his academic home for the next six decades. At Harvard, Meselson continued his study of DNA and RNA. He discovered an example in *E. coli* where the bacterium can use so-called restriction enzymes to cut and defeat an invading virus. Such restriction enzymes have since become a fundamental tool of genetic engineering, which typically involves the transfer of DNA from one organism into another. Meselson's work on restriction enzymes, first published in 1968, has had broad use in the field of genetics.

Work on Arms Control

In the aftermath of World War II, it became increasingly clear to many people that the power of the atom could be used to both serve and destroy humanity. As biologists like Meselson helped to unlock the secrets of genetics, they realized that the gene, like the atom, held enormous potential for both good and bad.

Meselson's interest in biological warfare began to develop in the early 1960s.[21] At Harvard, he attended a monthly arms control seminar for faculty, organized by a professor named Henry Kissinger. The two men, both going through divorces, became friendly.[22] In the summer of 1963, Meselson was invited to work in Washington, DC, as a consultant for the Arms Control and Disarmament Agency, an arm of the US State Department focused on the control of atomic weapons. "Scientists are arrogant enough to think they can contribute to anything," Meselson says.[23] He quickly realized that there were plenty of other people who knew more about nuclear weapons than he could learn in a summer, and he asked to instead to explore control of chemical and biological weapons, which was more familiar territory.

Meselson started his work with a tour of Fort Detrick in Maryland, home of the US biological warfare program. His guide pointed out a seven-story, windowless building. "That's where we make anthrax spores," Meselson remembers him saying in reference to the deadly bacteria. "This will save us a lot of money. It's a lot cheaper than nuclear weapons." The comment stuck with Meselson. "I don't know when exactly it went 'ping' in my brain," he says. But by the time he got back to his office, it occurred to

Biologists (left to right) Matthew Meselson, James Watson, and Walter Gilbert at Harvard University, ca. 1960s. *Source:* Courtesy, Cold Spring Harbor Laboratory Archives.

him that this idea didn't make any sense. "Would we want a ten-cent hydrogen bomb? What would the world be like if everybody could have one? Why would you want mass destruction to be cheaper than hydrogen and uranium and plutonium bombs? You wouldn't. It would be idiotic. In fact, it would be great if war was so expensive that nobody could afford it."

Meselson began a lobbying effort to stop the use of these weapons. His first contact was McGeorge Bundy, national security adviser to presidents John F. Kennedy and Lyndon B. Johnson. Bundy had hired Meselson while a dean at Harvard and, after hearing Meselson's argument, agreed to keep biological weapons out of US war plans. Another Harvard friend connected him to an editor at the *Washington Post*, who gave him contact information for science journalists across the country. One by one, Meselson reached out to the journalists, convincing most to write about the folly of pursuing such inexpensive weapons. This helped turn public opinion against biological and chemical weapons, says Meselson.

In 1967, during Johnson's presidency, Meselson and a colleague began a petition that drew the signatures of five thousand scientists, asking Johnson to review the US position on chemical and biological weapons and to ban the use of tear gas and herbicides in the Vietnam War. As far as Meselson is aware, no review was conducted at that time.[24]

Meselson realized that only attention from the president would change policy. When Richard M. Nixon was elected in 1968 and appointed Kissinger his national security adviser, Meselson sensed an opportunity. One day early in the Nixon presidency, Meselson literally collided with Kissinger on a ramp at Boston's Logan Airport. "My memory is I was going down and he was going up," Meselson says. Kissinger, knowing Meselson's interest in biological warfare asked, "What should we do about your thing?" Meselson offered to write a few papers laying out his argument.

The two men stayed in touch. At one point, Meselson sent Kissinger a copy of *The Andromeda Strain*, Michael Crichton's 1969 novel about an extraterrestrial microscopic organism that runs amok on earth. Kissinger enjoyed the medical thriller so much he passed it to the president, Meselson recalls, and it raised the president's consciousness about the havoc biology could cause.

Meselson says there were two key motivating factors behind his interest in biological weapons. One was that biology was his chosen subject matter. The other was his concern that war would turn into a nonstop event with no rules. Biological and chemical weapons were so inexpensive that countries didn't have to mount a large war effort to continue it indefinitely. "I had this feeling that maybe they had this capacity to change the nature of war completely," he says.

Kissinger initiated a review of the government's policy on chemical and biological weapons and with Meselson's backing convinced Nixon to support a new international treaty. The Biological Weapons Convention, initiated by the British in 1972, prohibited signatory countries not only from using biological and chemical weapons but from manufacturing and stockpiling them, which a previous international agreement did not, and it forbade their use as retaliation in addition to first use.[25] The US Senate ratified the convention in 1975. More than 180 countries are now party to the agreement. The convention has only rarely been violated in the more than forty-five years since it was approved.[26]

Meselson was also involved in efforts to ensure that the ban held. He was called on to explain an outbreak of anthrax in the Soviet Union that killed sixty-eight people in 1979. Allowed to investigate only from afar, Meselson found that the deaths had likely been caused by eating diseased cattle. In 1992, however, after years of effort, Meselson was finally allowed to lead a field investigation in the region. The research team, which included his wife, Jeanne Guillemin, a medical sociologist, was able to plot the location of victims' homes and workplaces against the direction the wind was blowing the day of the disaster. The map convinced Meselson that his previous finding was wrong. The team determined that the deaths resulted from an accidental release of anthrax spores from a Soviet biological weapons facility upwind of the affected area.[27]

In more recent years, Meselson has continued his work on genetic processes, including research on the asexual reproduction of a class of microorganisms known as *bdelloid rotifers*. Meanwhile, the genetics revolution that Meselson helped found has brought tangible benefits to humankind: new medical treatments, improved crop yields, a deeper understanding of our collective and individual ancestry, and a better understanding of the biochemical underpinnings of life. It has also brought worries about the consequences, both intended and unintended, of genetically modified organisms in our food supply and the possibility of superbugs escaping a lab and causing disease epidemics. Many ethical and moral questions emerge from the use of genetic knowledge. Among a handful of biologists of his generation, Meselson accepted the responsibilities that came with his discoveries.

RECOMBINANT DNA AND THE BIOSAFETY DEBATE

By the summer of 1976, after years of town-gown tensions, Cambridge Mayor Alfred Vellucci was fed up with Harvard and its scientists. He had heard the university planned to construct a new laboratory to do research on recombinant DNA—the process of splicing genetic material from one species into another—but he said that no

Mayor Alfred Vellucci speaking at a hearing on proposed recombinant DNA research in Cambridge, 1976. *Source:* MIT Infinite History website.

one from Harvard had briefed him on its plans.[28] Experimenting with the blueprints of life could combat disease and reduce global hunger by improving agricultural yields, scientists said. But Vellucci worried it could also expose his Cambridge constituents to unforeseen dangers. He argued that work should stop until his citizens' safety could be guaranteed.

For the previous two years, the federal government had frozen recombinant DNA research in labs it funded as it studied how to conduct the research safely. On June 23, 1976, the National Institutes of Health (NIH) unveiled new restrictions, based in part on the advice of MIT biologist David Baltimore. For the highest level of risk, research would need to take place within sealed chambers with elaborate ventilation systems, airlocks, and special clothing. The regulations largely resolved the concerns of scientists and government regulators.

In Cambridge, however, the debate was only beginning. The same day the NIH guidelines were published, Vellucci took a decidedly combative approach at the city's first public hearing on recombinant DNA experimentation.[29] He worried aloud that the researchers "may come up with a disease that can't be cured—even a monster." Vellucci instructed the scientists planning to speak at the hearing to avoid jargon: "Refrain from using the alphabet," he told them. "Most of us in this room, including myself, are lay people. We don't understand your alphabet, so you will spell it out for us, so

Harvard biology department faculty and students outside the university's Biological Laboratories, early 1960s. Among those pictured are molecular biology pioneers Mario Capecchi, Walter Gilbert, Alfred Goldberg, Alfred Tissières, and James Watson. Watson, Gilbert, and Capecchi would go on to win Nobel Prizes for their work. *Source:* Courtesy, Cold Spring Harbor Laboratory Archives.

we know exactly what you are talking about because we are here to listen." After the first few speakers, Vellucci interrupted with a series of questions: "Can you make an absolute one hundred percent guarantee that there is no possible risk which might arise from this experimentation?" "Is it true that in the history of science mistakes have been made or known to happen?" "Do scientists ever exercise poor judgment?" Vellucci proposed a two-year ban on all recombinant DNA experiments in the city, sparking an outcry from several scientists. Harvard biologist Mark Ptashne told Vellucci that a moratorium in Cambridge would derail the work of half of the school's biology department, "including experiments that no one sir, *no one*, has ever claimed had the slightest danger in them."

The city council rejected Vellucci's proposed two-year ban, but the debate continued. A few weeks later, at a second hearing, "visitors spilled over from the [council] chambers into the halls," according to Tufts University environmental policy professor

Sheldon Krimsky.[30] The scientific community was accustomed to governing its own affairs, Krimsky wrote, not to the type of public attention or local regulation that the genetics research issue had engendered in Cambridge. The city council passed a three-month moratorium and created a nine-member citizen review board to investigate the safety procedures of DNA labs in the city.[31]

According to Krimsky, who was named to that review board, the committee met in a small dining room in the Cambridge Hospital and held a "courtlike proceeding with scientists as advocates arguing their case before the citizens."[32] The board spent more than a hundred hours listening to testimony and deliberating. In January 1977, the board recommended creating a city committee to oversee recombinant DNA research and requiring that all genetic splicing experiments take place in specialized facilities, providing what is called P3-level containment.

The Cambridge City Council approved the committee's recommendations the following month, becoming the first municipality in the country to pass a biohazard containment law. All research with biological agents in the city had to be conducted following the National Institutes of Health biosafety guidelines, regardless of whether federal funds were used in the research. The city's regulations, which went beyond the NIH guidelines, also included regular site inspections and worker training.[33]

Some observers thought Cambridge was overreaching in its biosafety ordinance and would discourage genetics research from taking place within city limits. Paradoxically, the regulations had the opposite effect: they removed uncertainty for the universities and private companies that conducted recombinant DNA work, while reassuring residents that their safety would not be compromised.

Three years later, when Biogen—one of the first genetic engineering companies—was looking to move its headquarters to the United States, Cambridge became the obvious location. Harvard's Wally Gilbert and MIT's Phil Sharp were among the dozen scientists who had helped found the company. The pace of progress in Europe, where the company was initially based, was too slow, Gilbert says, so he and others wanted to move the company to the United States to speed things up.[34] In addition to being convenient for him and Sharp, Cambridge had the intellectual strength—with its nearby universities and hospitals—to propel progress. And "because we had had the debate about recombinant DNA in Cambridge," he says, it was the "natural thing to do," to locate the company here.

Biogen's arrival in 1982 was proof that the city was open to the fledgling biotechnology industry. Even Mayor Vellucci had had a change of heart about genetics research, reportedly remarking at a Biogen ribbon-cutting event that he had "no fear of recombinant DNA as long as it paid taxes."[35]

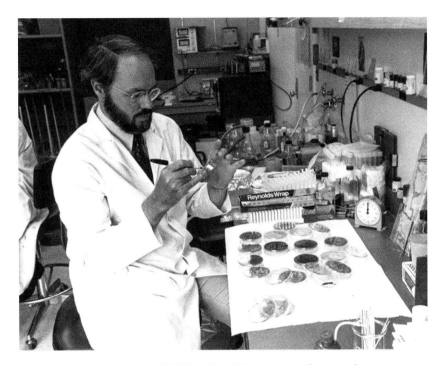

Phillip Sharp at work in his MIT lab, early 1980s. *Source:* Lasker Foundation.

A second wave of biotech entrepreneurs followed. A struggling start-up called Genzyme began manufacturing operations in Cambridge in 1986, under the direction of Henri Termeer, who transformed it into a major life sciences company with a specialty in treating rare inherited diseases.[36] Genzyme, which proved that treating rare diseases could be a viable business model, was acquired by French drug giant Sanofi in 2011. Termeer became a key player in the Cambridge biotech world, mentoring many of its leaders until his untimely death in 2017. A cluster of DNA research groups began to build in the formerly industrial Kendall Square area, drawn by the area's then-inexpensive real estate, empty factory buildings, and proximity to MIT, Harvard, and Massachusetts General Hospital, just across the Longfellow Bridge in Boston.

Today's Kendall Square, with its burgeoning skyline and buzzing street life, would be unrecognizable to a visitor from just a few decades earlier. An area that felt more like a cluster of empty lots is now almost built out. Surface parking lots have become research laboratories. The neighborhood's first grocery store does a brisk business.

Cambridge's biotech ecosystem has become the envy of cities around the world. Within a few square miles, ideas for new therapies are born in academic labs and nurtured at small firms funded by area venture capitalists. After showing early promise, some are purchased by large pharmaceutical companies which have R&D operations here and often tested in clinical trials at Boston hospitals.

In 2020, Takeda, a Japanese pharmaceutical company, was the city's fourth-largest employer, after Harvard, MIT, and the city government. The research arm of Novartis was fifth with more than 2,300 employees, closely followed by Biogen.[37] Other top ten employers included the Broad Institute, which focuses on genomic medicine, tech companies Google and HubSpot and a hospital.

Harvard Business School professor William Sahlman says that a number of factors conspired to nearly kill the region's biotechnology industry before it even got started.[38] In addition city leaders' skepticism about the safety of biomedical research, Harvard was not always supportive when faculty members started commercial businesses, Sahlman notes. Cautious Boston bankers were hesitant to lend money to risky biotech start-ups. The region's hospitals were slow to participate. It was the vision of a few well-connected scientists and venture capitalists with support from MIT who launched Cambridge's biotechnology innovation ecosystem, Sahlman asserts. "In the end," he says, the flywheel of this innovation system couldn't be held back, and "it all broke loose."

UNRAVELING THE HUMAN GENOME

On June 26, 2000, standing at a White House podium, President Bill Clinton announced the completion of the first detailed draft of the human genome, the sequence of all of the genes in the human body.[39] He compared it to a map hanging just down the hall that explorer Meriwether Lewis unveiled to President Thomas Jefferson two hundred years earlier. This new genetic map, Clinton said, was even better than the one that first depicted the American West. "Without a doubt this is the most important, most wondrous map ever produced by human-kind," Clinton enthused to a roomful of politicians and scientists. The international Human Genome Project, which was formally completed three years later, had already taken a decade and would end up costing about $2.7 billion in US funding,[40] plus more from private entities and other governments.

Biologist Francis Collins, who led the American government's effort, predicted that the project would be sufficient to "unravel the mysteries of human biology" and potentially lead to the end of cancer in his lifetime. According to a statement from the Clinton White House at the time, "Scientists will be able to use the working draft of the human genome to: alert patients that they are at risk for certain diseases; reliably predict the course of disease; precisely diagnose disease and ensure that the most effective

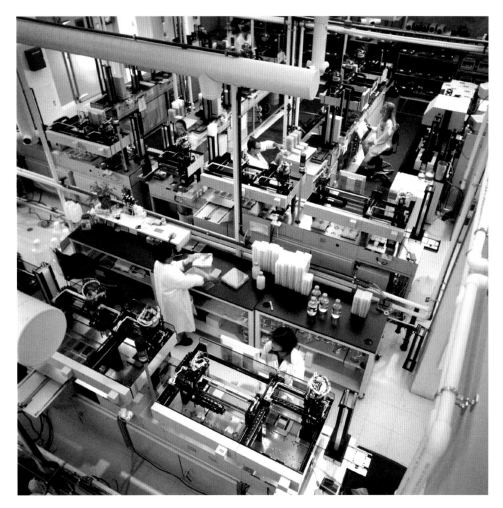

View of a genome sequencing laboratory at the Whitehead/MIT Center for Genome Research, ca. 2001. *Source:* Courtesy, Whitehead Institute.

treatment is used; and develop new, more effective treatments at the molecular level." Although human genetics has proven far more complicated than the scientists and politicians knew or acknowledged then, the map unveiled that summer day—and the multitude of genetic sequences that followed—have transformed the practice of medicine and many other facets of modern life.

Cambridge-based researchers played key roles in mapping the human genome and in experiments the project spawned. Scientists at MIT and Harvard helped generate the original idea of the project, shepherded it through a challenging political climate, and undertook fully a third of the genome sequencing—more than any other partner in the global consortium of laboratories that participated. Eight representatives of the MIT-affiliated Whitehead Institute attended the White House ceremony that day as an acknowledgment of their role in the work.[41]

"It's not an accident that the genome project got done in the MIT ecosystem," says Eric Lander, who led the Whitehead's effort and later became the first Cabinet-level scientific advisor to a US president.[42] MIT's strength in interdisciplinary research—with top scientists in biology, math, computer science, and engineering all located in the same place—meant that problems could be addressed as they arose, Lander says. The Whitehead, across the street from MIT and staffed by its faculty, was able to solve financial challenges that some of its peer institutions could not. The National Institutes of Health promised Lander $40 million to buy gene-sequencing equipment—but spread the funds across five fiscal years with no guarantee of money after the first. The Whitehead took out a $40 million loan and bought the equipment up front.

Factory buildings, emptied by the city's deindustrialization, were also available for rent in Kendall Square. The first home for the genome project's sequencers was a warehouse that had previously held concessions—beer and popcorn—destined for fans at Boston's Fenway Park.

Lander went on to found the Broad Institute of MIT and Harvard, a biomedical and genomic research center next door to the Whitehead. He points out that the Human Genome Project has brought tremendous spinoff benefits to the region's life sciences community. It opened the eyes of young students to research now possible using the genome sequence and by 2020, Lander says, had led to more than sixty collaborations across Boston in genomic medicine.

While in many places the urge to maintain the status quo can stifle innovation, Lander says, the annual influx of students to the region provides a constant flow of new ideas. "Around here, people are both edgy—they are aware that what worked in previous generations isn't going to work in the next generation—but [they also think]

President Bill Clinton congratulates Francis Collins and J. Craig Venter at a White House event announcing the completion of the first survey of the entire human genome, June 26, 2000. Eric Lander is in the foreground. *Source:* Courtesy, William J. Clinton Presidential Library; White House Television (WHTV), "Announcing the Completion of the First Survey of the Entire Human Genome at the White House," Clinton Digital Library, accessed December 3, 2020, https://clinton.presidentiallibraries.us/items/show/16115.

you might as well take intellectual and other assets built in the previous generation and use them as jumping off point for the next."

In one sense, the Human Genome Project can trace its origins to Cambridge, England, where in 1953 Francis Crick and James Watson first described the structure of DNA based on images taken and calculations made by Rosalind Franklin. The genome, they found, is essentially a very long string of chemical compounds arranged into the form of a double helix—something like a twisted stepladder. Four nitrogen-based compounds (adenine, guanine, cytosine, and thymine) comprise the genetic alphabet of DNA (deoxyribonucleic acid). Sequences of the four compounds (abbreviated as A, G, C, and T) create the equivalent of words and phrases. This genetic literature provides the blueprint for all the activities of every living cell.

The Human Genome Project also traces its origins to the work of MIT biologist and yeast geneticist David Botstein, who in the 1970s began writing a paper detailing an idea for mapping the genome.[43] Although the Human Genome Project wasn't given

The Whitehead Institute on Main Street in Cambridge. The Whitehead/MIT Center for Genome Research undertook sequencing of the human genome beginning in 1990. *Source:* Photo by the authors.

Eric Lander at the Whitehead Institute. *Source:* Courtesy, Whitehead Institute.

that name until 1986, Botstein's paper, published in 1980, called for the construction of a "genetic linkage map." "It was essentially the first call to arms—the first call to action to thinking about genetics in a systematic way," says biologist David Page, who worked in Botstein's lab at the time and would go on to run the Whitehead Institute from 2004 to 2020.

The idea for a map captured the imagination of some biologists and infuriated others, who thought that such a monumental project would be an expensive waste of effort and divert resources from other priorities. The naysayers could easily see themselves as realists. "Back in 1985, a linear estimate would say that it would take two hundred years of work worldwide to completely sequence the human genome," Lander says.

Harvard biologist Wally Gilbert, who shared the 1980 Nobel Prize for developing an early gene-sequencing technology, became a champion of the idea. In the summer of 1985, Gilbert says he was invited to a meeting at the University of California, Davis, about sequencing the entire human genome.[44] "I go to the meeting thinking that's the silliest idea I've ever heard," he says. But at the meeting, learning the rate of sequencing that was then possible, he "became convinced that it was a perfectly technologically feasible industrial problem." On the plane back, he drafted a letter to the meeting's organizer, outlining how a sequencing center could map the entire genome by hiring about three hundred people to work for thirty years. "You'll do 1 percent the first decade, you'll do 10 percent the next, and the last 90 percent the decade after that, because the technology will steadily improve," he remembers writing. He predicted the entire venture would cost about $3 billion—which turned out to be pretty accurate. Gilbert wrote enthusiastically at the time: "The total human sequence is the Grail of human genetics—all possible information about the human structure is revealed (but not understood). It would be an incomparable tool for the investigation of every aspect of human function."[45]

To read the entire human genetic code, the Human Genome Project first broke the long string of DNA into separate fragments whose position was known. This is like dividing a book into chapters and noting their order. Each of the fragments of genetic material was then sequenced, its individual chemical letters decoded using robotic machines linked to computers. The book was then put back together by reassembling individual fragments, noting where they overlapped.

Decoding the genes that make us human relied on advances in molecular biology, chemical engineering, and computing. "The Human Genome Project is the direct descendant of the whole unexpected confluence of genetics and information theory," Maynard Olson, a geneticist at the University of Washington, wrote in 2002.[46] This convergence "must rank as one of the great coincidences in the history of science and technology," Olson wrote, referencing the work of Claude Shannon (see chapter 6) that allowed all information to be encoded in digital form. "In the same historical instant, humans discovered that biological information is digital . . . and, quite independently, invented new technological means of storing, processing, and transmitting information based on digital codes."

The Human Genome Project officially kicked off in 1990 and published the first draft of the genome in the journal *Nature* in February 2001.[47] The majority of the sequencing work occurred in the nine months prior to Clinton's 2000 announcement, with machines recording "one thousand letters of the DNA code per second, seven days a week, 24 hours a day."[48] The effort to crack the code involved sixteen laboratories in the United States, United Kingdom, Japan, France, Germany, and China. A near complete version was finished in 2003 and published in 2004.[49]

Today, instead of taking fifteen years and $3 billion to roughly sequence a single human genome, Lander notes that the Broad Institute sequences many human genomes in parallel, with a far more accurate sequence completed roughly every ten minutes and for less than $650. Private companies claim to do the same for $100.[50]

Lasting Impact

It's hard to overestimate the impact that genetic sequencing has had on our daily lives. A deeper understanding of the genetics of cystic fibrosis, for instance, has helped Boston-based Vertex Pharmaceuticals develop treatments for the vast majority of patients with the disease.[51] Cancer care has also made tangible use of genetic sequencing techniques. Major cancer centers in the United States routinely sequence patients' tumors to help determine their course of care.

Genetic sequencing technology has also transformed the criminal justice system. After the terror attacks in New York City on September 11, 2001, investigators used DNA analysis to identify the remains of victims.[52] Murderers are tracked down by the cells they unintentionally leave at the scene of their crime. The so-called Golden State Killer was identified in 2018 via distant cousins who volunteered their genetic information on a genealogy website.[53]

The human genome is also teaching us about ourselves, revealing great migrations. Scientists, like David Reich at Harvard,[54] are now reconstructing the movements of ancient people by comparing the DNA of bodies discovered in caves and atop frozen mountains to their descendants' genomes.

In addition to reading the genome, scientists are learning to write it, altering human genes to cure disease, plant genes to improve crop yield, and animal genes for everything from medical research to supposedly healthier meat to species conservation.

Another measure of the Human Genome Project's lasting influence: it's hard to imagine not having it. Lander says that at a lab meeting around 2018, one of his graduate students stopped midsentence, turned to him and said: "before the Human Genome Project, how did you ever get anything done?" The project that took so many years of effort and so much money is now taken for granted, the way we take electricity or the internet for granted. "Success," Lander says, "is that people assume the Human Genome Project was always there—or was done in the 1950s or 1840s, because you can't imagine a world without it."

ROBERT LANGER: BIOMEDICAL ENGINEERING PIONEER

On March 16, 2020, four people in Seattle bared their upper arms and allowed a needle to pierce their skin, receiving the first doses of a vaccine that would change the course of the worldwide COVID-19 pandemic, the worst global disease outbreak in a century. The genetic sequence of the novel SARS-CoV-2 virus on which the vaccine was based had been publicly released only sixty-four days earlier, on January 11. No vaccine had ever been developed in such a short period of time. "There's no doubt that that's the world indoor record," Anthony Fauci, the head of the National Institute of Allergy and Infectious Diseases and a vaccine expert, said later.[55] "I've never seen anything go that fast."

Fauci's staff had helped tremendously with the development of the vaccine, figuring out before the 2020 pandemic started that a coronavirus vaccine should target the spike protein on the virus's surface and be configured in a specific way. Although many COVID-19 vaccine developers used the same target, the company Fauci's staff collaborated with most closely—the one that produced the vaccine shot into arms that Monday in Seattle—was a ten-year-old Cambridge biotechnology firm called Moderna.

It was founded with the idea of directing the body's cellular machinery to manufacture specific proteins that could treat disease or trigger an immune response, spurring the body to produce the antibodies that would protect it from an invading pathogen. The vehicle Moderna used to control the cellular processes was something called messenger RNA (mRNA)—the intermediary between an organism's DNA blueprint and the proteins that direct all the actions required for life. Moderna scientists were developing vaccines to protect against the flu and various childhood diseases. They had also been working on personalized cancer vaccines, hoping to turn the body's own immune system against its tumors. And they had begun to address rare diseases, helping the body make proteins that were missing because of a genetic fluke.

The company's four founders represented the region's diverse biotech ecosystem: a Cambridge-based venture capitalist, a cardiovascular expert at Massachusetts General Hospital, a Harvard Stem Cell Institute researcher, and an MIT biomedical engineer named Robert Langer. The group's audacious goals and the reputations of its founders sparked great enthusiasm on Wall Street. Although Moderna hadn't yet brought a product to market, its initial public offering in 2018 was then the largest in the history of biotech, putting the value of the company at nearly $8 billion.

Langer, the biomedical engineer, was already quite familiar with success. He has had an astonishingly productive career at the intersection of chemical engineering and medicine. As of the time of this writing, Langer has helped create forty-three companies, forty-one of which either are still around or have been bought by a larger company.[56] He has mentored somewhere around a thousand graduate and postdoctoral students. For his contributions, Langer has received hundreds of awards, including the United States National Medal of Science and the National Medal of Technology and Innovation.[57] He is one of the few people named to all three National Academies: Sciences, Engineering, and Medicine. And he's one of ten who currently hold the post of Institute Professor at MIT, the school's highest teaching honor.

Robert Langer in his MIT laboratory. *Source:* Photo by Ben
Tang, courtesy, Langer Research Laboratory, Massachusetts
Institute of Technology.

Nothing in Langer's early life particularly foretold his later path. He grew up in
Albany, New York, the older of two children of a liquor store owner and a homemaker.[58] His father and grandfather played math games with him as a boy. Langer
remembers having fun with his Gilbert erector set, his microscope, and especially a
chemistry set. He set up a laboratory in his basement where he played with chemical reactions and made rubber. He didn't know what he wanted to do with his life,
but because of his facility with math and science, his family encouraged him to study
engineering in college.[59] At Cornell University, his favorite class was chemistry, so he
majored in chemical engineering. He moved to MIT to pursue a graduate degree, still
unsure of what career he wanted to pursue.

Langer received his doctorate in chemical engineering in 1974 just after the Arab
oil embargo lifted. Like his MIT classmates, he was inundated with job offers from

Robert Langer at Cornell University, where he received a degree in chemical engineering in 1970.
Source: Courtesy, Langer Research Laboratory, Massachusetts Institute of Technology.

oil companies: twenty overall, four from Exxon alone.[60] In one interview, in Baton Rouge, Louisiana, an Exxon engineer told him that if he could help increase the yield of just one petrochemical by 0.1 percent, it would be worth billions of dollars. "I remember flying back to Boston that night thinking I really didn't want to do that." He began searching for a way to make a larger contribution in the world.

After months hunting unsuccessfully for a job, at a friend's recommendation, Langer found himself meeting with Judah Folkman, a doctor at Boston Children's Hospital. Folkman had a novel idea for fighting cancer but was struggling to implement it. He knew there was a mechanism that could stop cancerous tumors from building their own blood supply, but the biologists and doctors he'd hired hadn't been able to figure out how to control it.[61] "So he thought, when he looked at my resume, and talked to me . . . maybe if that person thinks differently, he'll be able to solve it," Langer said.[62] And he did.

Langer purified compounds from slaughterhouse waste, testing them for their effectiveness in fighting cancer.[63] He then developed special polymers that would slowly release the cancer-fighting compounds so they could be delivered at a steady, measurable rate.[64] Addressing this problem was "a wonderful opportunity to do something important for really the first time in my life," he told a radio interviewer in 2019.[65]

Finding a method for cutting off a tumor's blood supply didn't provide a cure for all cancers, as Folkman had hoped, and it took twenty-eight years to bring to market a product based on what's called angiogenesis inhibition. But today, drugs that build on Folkman's idea—including Avastin, which brought drug company Roche $48 billion in sales between 2004 and 2019[66]—are used to treat many types of cancer,[67] as well as major causes of blindness among older adults. In Folkman's lab, Langer developed the first approaches for delivering proteins as well as DNA and RNA into the body. Decades later, this provided the underpinnings for the methods used by Moderna and other companies to encapsulate and protect the fragile messenger RNA that is the basis for several COVID-19 vaccines.

The headquarters of Moderna in Cambridge's Technology Square in 2020, when the biotechnology firm gained widespread acclaim as the developer of a successful COVID-19 vaccine. Robert Langer was a Moderna cofounder, early investor, and board member. *Source:* Photo by the authors.

In 1984, Langer and neurosurgeon Henry Brem of Johns Hopkins University in Baltimore developed a plastic wafer that could be impregnated with chemotherapy drugs and slowly released cancer-fighting medicine into the brain. The biodegradable wafer, called Gliadel, was approved in 1996 to treat glioblastoma, an aggressive brain cancer.[68] Shown to extend the lives of patients with this fast-moving, usually fatal cancer, it remains in use today.[69]

In the mid-1980s, a researcher at Boston Children's Hospital approached Langer with an idea for creating engineered human tissue by creating a microscopic scaffold to hold cells in place.[70] Their 1993 paper[71] helped define the field of tissue engineering and led to functional skin, bones, blood vessels, and cartilage.[72] It is also used to simulate human organs in a dish, as a complement to animal research.

His work in tissue engineering reflects Langer's broader approach to science, says Ali Khademhosseini, a biomedical engineer and entrepreneur who completed his postdoctoral studies with Langer.[73] "His lab is always at the cutting edge. They're never satisfied. They're always creating new areas," Khademhosseini says. "If it's been done before, then it's not worth even going in that general direction. He's always trying to be first in a whole new area."

One of Langer's experiments with implantable drugs involved the creation of microchips, much like those used in electronic devices, that, when surgically implanted in the body, can be activated remotely to release tiny amounts of medicine over the course of months or years. In 1999, Langer and MIT colleague Michael Cima started a small biotech company called Microchips Biotech to commercialize the concept, which might prove applicable to an entire range of chronic medical conditions requiring regular small doses of drugs. Clinical trials have begun in people. One of the company's products, funded in part by the Bill and Melinda Gates Foundation, is intended to deliver contraceptives and can be turned on and off for family planning.[74]

Langer credits his students and peers with helping him remain innovative. Cambridge has a remarkable entrepreneurial spirit, with Kendall Square housing the highest concentration of biotech companies in the world. "For a biomedical engineer," he says, "this has to be the best place in the world to be."

GENE-EDITING TOOLS

On February 4, 2011, a month after joining the faculties of MIT and the Cambridge-based Broad Institute, Feng Zhang was sitting in the back of a conference room, listening to a lecture on pathogenic bacteria.[75] He was intrigued by mention of a strange acronym the speaker casually mentioned: "CRISPR." Curious, Zhang turned to Google to find out more and learned that CRISPR (clustered regularly interspaced short palindromic repeats) is an immune system from bacteria that recognizes and cuts the DNA of invading viruses. Zhang had spent the previous year as a junior fellow in a Harvard research lab, working with complicated, cumbersome gene-editing tools that cut DNA. He was immediately interested in turning CRISPR into a gene-editing tool, he says.

A day later, he flew to Miami to attend a research conference, where he spent most of his time holed up in a hotel room reading the few research papers that had been written about this bacterial immune system.[76] Several were by scientists in the food industry who were interested in using CRISPR to improve the taste of yogurt by getting rid of unwanted viruses. But Zhang, at age twenty-nine, the youngest scientist ever given his own lab at the Broad Institute, had other ideas. If he could harness CRISPR—essentially molecular scissors—to work in human cells instead of bacteria, that might get him one step closer to his dream of reversing neurological conditions like autism, depression, and schizophrenia.

From his Miami hotel room, he emailed his doctoral student Le Cong about CRISPR, suggesting that "This could be really big." Within days of his return from Miami, Zhang and Cong had adapted the bacterial components for use in human cells and demonstrated their first success. The pair continued to optimize this CRISPR system to robustly edit specific DNA sequences in cells with nuclei.

Zhang's interest in CRISPR connects him to a long line of Cambridge-based pioneers in the field of gene-editing tools. Matthew Meselson had used restriction enzymes to cut DNA in the late 1960s at Harvard. Meselson's biology department colleague Wally Gilbert developed one of the first two gene-sequencing methods. In the 1990s and 2000s, George Church, who did his doctoral studies in Gilbert's lab, became the driving force behind more than twenty companies developing DNA sequencing technologies. Church's Harvard Medical School lab is in Boston, but he considers himself part of the Cambridge biotech ecosystem and regularly teaches at Harvard and MIT.

Zhang had recently finished doing postdoctoral work with Church when both men's labs, independently, began to pursue CRISPR gene editing in mammalian cells. In August 2012, the journal *Science* published a blockbuster paper that transformed the quiet field of CRISPR research.[77] Jennifer Doudna at the University of California, Berkeley—who had earned her PhD at Harvard in the same department where Church worked—and Emmanuelle Charpentier, a French researcher, showed they could use CRISPR and a related enzyme called Cas9 to precisely edit the genes of bacteria.

Zhang submitted his own paper to *Science* showing he could use CRISPR/Cas9 to edit mammalian cells. The journal published both his paper and a similar one from

Feng Zhang of the Broad Institute. *Source:* Justin Knight, McGovern Institute.

Church on January 3, 2014. "Getting the system to work in human cells was a major step forward in developing a new type of genetic therapy," science writer Kevin Davies writes in a 2020 book on gene editing.[78]

    The new tool had a rapid and profound effect. In awarding Doudna and Charpentier the 2020 Nobel Prize in chemistry, committee member Pernilla Wittung-Stafshede said, "The ability to cut DNA where you want has revolutionized the life sciences.[79] The 'genetic scissors' were discovered just eight years ago, but have already benefitted humankind greatly." Laboratory researchers quickly took up CRISPR/Cas9 as easier

George Church in 2018, by Heytessa. Licensed under
CC BY-SA 4.0 (https://creativecommons.org/licenses
/by-sa/4.0/deed.en).

to use than its two predecessor gene-editing tools. Labs began editing mouse genes one
by one to see what happened. Others experimented with ways to improve agriculture
by knocking out and then adding genes using CRISPR.

By late 2020, CRISPR gene editing had begun to be used for treating—and potentially curing—a variety of inherited conditions, beginning with eye diseases and blood
disorders like sickle cell disease. Testing the method, researchers from two Boston-area
companies, Vertex Pharmaceuticals and CRISPR Therapeutics, gene edited the blood
of one patient with sickle cell disease and another with a disorder called beta thalas-
semia. More than a year later, the scientists showed their edits were effective and had
changed the course of the two patients' diseases.[80] Both had been able to stop regular
blood transfusions, and the person with sickle cell hadn't had any pain crises in the year
since their one-time treatment.

Facile gene editing also has potential pitfalls. Some fear that gene editing will be used to create "designer babies" whose genes are manipulated not to avoid disease, as most scientists intend, but to boost attributes like intelligence and appearance.

Next-Generation Gene Editing

In 2013, David Liu, who has labs at both the Broad Institute and Harvard's chemistry department, realized one flaw with CRISPR. While its scissors could make cuts in a precise spot, they sometimes led to small insertions and deletions at the cut site, as well as other undesired changes. "CRISPR as it exists in nature and was originally used was not evolved to fix anything. It was evolved to destroy DNA," Liu says.[81] "For the vast majority of known gene mutations that cause disease, it's difficult to understand how destroying the gene is going to benefit patients."

In late 2013, Liu began trading emails about this problem with a California woman, Alexis Komor, who had just accepted a postdoctoral position in his lab. She arrived in Cambridge in early 2014 and immediately set to work designing a more precise gene-editing tool—what Liu would later describe as a pencil instead of a pair of scissors.

David Liu of the Broad Institute. *Source:* Casey Atkins Photography, courtesy of Broad Institute.

In the four-letter DNA alphabet, C (representing the nucleotide cytosine) always binds to G (guanine) and A (adenine) to T (thymine)—or, in RNA, to U (uracil). Komor and Liu identified a natural enzyme from rats that chemically changes the letter C to a U. They stitched it onto a disabled version of CRISPR that could no longer cut DNA. The result was an engineered protein that converted a target C to a U in a test tube. But in cells, the DNA repair system would quickly cut out the U and change it back. Komor spent months trying to figure out how to preserve the correction. Finally, while talking to a colleague in the lab's kitchen, inspiration struck.[82] She realized that if she nicked the opposite strand from the one she had just changed, the DNA repair system would probably replace it—matching her U with an A and maintaining her fix.

Komor ran into Liu in a hallway one weekend day and shared her insights. "I think I have a really good idea for how to get that mismatch problem to be resolved in our way," he remembers her telling him. "The beauty of that idea was a very, very simple way to solve the problem." She had already tinkered with the CRISPR-Cas9 "scissors," making two alterations that enabled her to target DNA letters precisely but not to cut either DNA strand. By undoing one of those, she was able to nick the opposite strand without cutting the double helix completely. Komor had developed the "pencil" Liu wanted, publishing her results in April 2016.[83]

Komor's base editor, called based editor 3 (BE3), can be used to fix only an estimated 14 percent of the single-letter genetic errors that cause disease—the ones in which a C-G needs to be changed to a T-A.

It took another member of Liu's lab, a graduate student named Nicole Gaudelli, to fix another 48 percent. Liu had suggested she try to make an A-to-G editor. This was tricky because there was no natural enzyme that could do this—so she had to make one. Starting with an enzyme that performed related chemistry on RNA, she began trying to "evolve" the missing enzyme. She set up Darwin-inspired natural selection in the lab, making hundreds of millions of random variants of the natural enzyme and rewarding those that performed the desired reaction. Each successive enzyme was better in some ways but not good enough in others. Six times, she went back to her lab plates to spend weeks trying to evolve an enzyme that had all the properties she wanted. Finally, after seven rounds of evolution, Gaudelli found her A-to-G editor, called ABE.[84] This was a pencil that could be used to erase thousands of disease-causing genetic mistakes.

One of the first diseases Liu tackled with this pencil was progeria, an extremely rare genetic condition that causes rapid aging. A six-year-old with progeria might develop heart disease normally affecting senior citizens, and children with this genetic mutation die on average at age fourteen. At the time, Liu had a new graduate student, Luke

Koblan, whose undergraduate studies at Harvard had given him connections at a local research lab. They allowed him access to human cells from children with progeria.

Koblan used Gaudelli's A base editor to fix the genetic error in 90 percent of the cells in a petri dish. Then the team moved its gene-editing tool into mice, more than doubling the animals' lifespan from 215 to 510 days—to the cusp of normal old age.[85] Treating a mouse is not the same as treating a person, but as of this writing, Liu and his collaborators are taking steps to bring this approach to patients.

Liu's lab has since developed an even more advanced form of gene editing, called prime editing, that promises to allow edits of most of the remaining diseases caused by single-gene mutations.[86] Zhang and Church, too, are working on improvements that will make gene editing more accurate, versatile, and cheaper. Church says a "huge fraction" of the innovations that have moved gene editing forward happened in Cambridge labs—"bam, bam, bam—and there was almost no competition (from labs elsewhere in the world)."

All of this gene-editing innovation has led to the creation of a half dozen biotech companies in Cambridge, most cofounded by Church, Zhang, and/or Liu. Charpentier, living in Berlin, cofounded CRISPR Therapeutics, which has headquarters in Zug, Switzerland, and Cambridge. Doudna also cofounded two Cambridge-based companies. Kevin Davies describes Doudna as "a mini-ecosystem in and of herself" and says she might help California eventually threaten Cambridge's supremacy in gene-editing companies.[87] "I wouldn't feel confident to say the future of gene editing is inevitably going to be in Boston," he says, though he adds that the presence of Zhang, Liu, Church, and others certainly argues for it.

While scientific advances occur all over the world, Liu describes the Cambridge-Boston area as "an intellectual Venice of the modern age." Part of the fabric of living in the region, he says, is that casual interactions are an everyday occurrence. Most of his lab's research projects, Liu says, came about by having chance encounters in a hallway or over a coffee pot or by hearing a lecture from one of the world-class scientists who speak in the area nearly every day. "I think that's precisely why it's become such a nexus of activity and productivity in academia, in biotech, and in healthcare," Liu says . "The overlap of those three segments is what powers many people's activities including our own research groups.

For a city known for its bookworms, scientists, and political activists, Cambridge has had a surprising and significant impact on American popular culture. Cambridge's young vibe, diverse communities, and openness to new ideas have long attracted creative people, some of whom have managed to become national and occasionally global figures. The city's academic institutions have a strong influence on its creative culture, attracting and nurturing emerging talents, but many people without direct connections to Cambridge's universities have been drawn into its lively cultural environment.

Harvard Square played host to a flourishing folk and rock music scene in the 1960s and 1970s. Political activist and folk singer Joan Baez gave some of her first public performances in 1958 at a music venue on Mount Auburn Street called Club 47. She sang twice a week, earning $25 a night.[1] She released her first album, the self-titled *Joan Baez*, to wide acclaim two years later. In 1961, Bob Dylan, then a twenty-year-old from Minnesota, began performing in Harvard Square, absorbing the local folk scene while he honed his craft. A year later, he released his first album, consisting mostly of traditional folk and blues tunes and titled simply *Bob Dylan*. Both Baez and Dylan would make it big in New York's Greenwich Village, but their early experiences in Cambridge helped shape their music and on-stage personas.[2] Van Morrison, the Irish singer-songwriter, wrote the words to *Astral Weeks* on Green Street near Central Square in 1968. The album is often considered among the greatest rock records of the twentieth century.

In 1974, Boston music reviewer Jon Landau was wowed by a little-known performer he saw opening for Bonnie Raitt at the Harvard Square Theater. "I saw rock and roll future and its name is Bruce Springsteen," he wrote in the *Real Paper*.[3] "And on a night when I needed to feel young, he made me feel like I was hearing music for the very first time." Springsteen's *Born to Run* album received rave reviews the following year.

The constant influx of new people looking for their own sounds, voices, and identities has continued to drive the local arts scene. Cambridge's relative compactness and dense neighborhoods often foster strong social connections. We highlight a few examples of Cambridge's influence on America popular culture in this chapter.

The **first football game** that resembled the modern American-style matchup, it can be argued, took place in Cambridge. In 1874, McGill University athletes challenged Harvard men to play a game they'd never tried before, more like English rugby than the soccer-style matches they were accustomed to. The game was a crowd pleaser. Though the Harvard faculty thought the sport was dangerous and a distraction to students—and even banned it briefly in the 1880s—intercollegiate football only grew in popularity.

In 1961, Paul and **Julia Child** moved from Paris to Cambridge at the end of Paul's career in the foreign service and following the publication of Julia's first book, *Mastering the Art of French Cooking*. The couple were drawn here via Julia's friendship with Avis DeVoto, a book editor and Harvard faculty spouse. Child's expertise soon earned her a spot on a local television station. During the forty years she lived in Cambridge, Julia Child became a staple of public broadcasting and an American cultural icon.

Benjamin Thompson came to Cambridge in 1945 to start a design firm with modern architecture giant Walter Gropius, founder of the Bauhaus and then head of Harvard's architecture department. Together with several partners, they formed an architecture practice that helped bring European modernism to America. In the 1950s, Thompson launched a fabled housewares store in Harvard Square called Design Research and later opened Harvest, a popular Cambridge restaurant. In renovating Boston's Quincy Market into a retail complex during the 1970s, Thompson helped create the "festival marketplace" urban development concept with the intent of drawing suburbanites and tourists back to aging city centers.

Cambridge appears in a number of well-known Hollywood movies. It seemed natural for scriptwriter and classics professor Erich Segal to set *Love Story* in Cambridge because he knew it well from his years studying and teaching at Harvard. Childhood friends Ben Affleck and Matt Damon performed in plays at Cambridge's public high school before they cowrote the screenplay to *Good Will Hunting* (1997) and moved to Los Angeles to pursue acting careers. Other popular films set in Cambridge include *The Paper Chase* (1973), *Legally Blonde* (2001), and *The Social Network* (2010).

Pediatrician **T. Berry Brazelton** became a trusted adviser to several generations of American parents as the author of two dozen books on childcare and from 1983 to 1995, the host of a cable television program called *What Every Baby Knows*. He simultaneously tended to the colds, cuts, and aches of thousands of children at his Cambridge pediatric practice and Boston Children's Hospital.

What Julia Child did for French cuisine and public television, brothers Tom and Ray Magliozzi did for automotive repair and public radio. Child employed her winning personality in convincing viewers that no recipe was beyond reach; the Magliozzis, hosts of *Car Talk* on National Public Radio (NPR) beginning in 1987, demystified car repair. At the same time, they broadened the radio network's popular appeal through their humor and infectious laughter. The brothers grew up in Cambridge and both attended MIT, and they often made reference to their home town ("our fair city," as they called it) in their weekly broadcasts.

The Harvard-educated cellist **Yo–Yo Ma** has performed and shared his love of music all over the world but has made Cambridge his home base for decades. Ma, a master of European classical music, has integrated musical traditions from across the globe into his repertoire.

THE ORIGINS OF MODERN FOOTBALL

On May 15, 1874, the McGill University football team had the advantage of speed: three of its players were among the fastest sprinters of their time.[4] Harvard had a home-team edge, playing on the school's Jarvis Field in Cambridge. But Harvard had never played by the Canadians' rules or with their oval-shaped ball, and neither team dominated the game. "The tide of battle surges up and down the field," according to an account of the contest.[5] The assembled crowd, numbering in the hundreds, cheered as each team advanced, thrilled by the close competition. Teammates and spectators agreed that this match was far more fun than the soccer-like match the two teams had played the day before. "The rugby game is in much better favor," a Harvard student journalist wrote, "than the somewhat sleepy game played by our men."[6]

By some measures, the matchup that day between Harvard and McGill was the first "real" game of American football.

McGill's team captain David Rodger had challenged Harvard to the competition—offering one game played on each campus. But the Harvard players were forbidden from leaving school during the term, so both games were held on Jarvis Field—now the site of Harvard Law School dorms—on Thursday, May 14, and Friday, May 15, 1874.

The first intercollegiate match of soccer-style football had been played five years earlier between Rutgers and Princeton using a round ball. It was this style of play that Harvard and McGill employed in their Thursday game. Harvard scored three goals in twenty-two minutes in the sweltering 85 degree heat; McGill none.[7] The *Harvard Crimson* newspaper wrote that the home team "won the three goals so easily that the McGill players seemed standing in the field merely to be spectators of their opponents' excellent kicking."[8]

The next day, the rules were switched. The two played a version of English rugby, with a large leather, oval-shaped ball.[9] The spectators, mostly fellow students, had paid fifty cents apiece to watch the matchup, with the second game even better attended

Iasigi   Gray   Watson
Prince   Faucon   Cate Morse   Ellis, *Capt.*
Sanger   Wetherbee   Whiting

HARVARD *vs.* McGILL

Harvard playing McGill in football at the Montreal Cricket Grounds under McGill's rugby-like rules, October 23, 1874. *Source:* Courtesy, Harvard Athletics.

than the first.[10] The earnings were devoted to the traveling team's transportation and to a festive dinner with free-flowing champagne.[11]

The claim to "first football game" remains contested. Princeton and Rutgers argue for their 1869 match. (Rutgers won 6–4.)[12] But the Harvard-McGill rules were far closer to what we know today as American football. Under McGill's rules, the ball had to be kicked over a ten-foot-high cross bar supported by two poles—which cost Harvard $2.50 to build.[13] The side that caught the ball could run with it until the player decided to kick it or the ball was forced from him.[14]

Tufts University in nearby Medford, Massachusetts, also claims a piece of football's origin story. Its game with Harvard, played under the rugby rules on June 4, 1875, on Jarvis Field, was what Tufts asserts was the first true intercollegiate football game played between US teams. The Harvard men had new uniforms, including white shirts, crimson trimmed pants, and crimson stockings.[15] Tufts won.[16]

In the fall of that year, Harvard challenged Yale to a game under the Rugby Union Rules, similar to those they'd played against McGill. Yale accepted, though it modified the rules some more. The teams played on Yale's home turf in New Haven, Connecticut. "The Harvard-Yale game was the pivotal point in the downfall of soccer and the rise of rugby in America," one author argues.[17] "As the younger and still less prestigious college, Yale had little choice [but] to play Harvard's game if it wanted Harvard's rivalry,"[18] Mark F. Bernstein writes in his book *Football: The Ivy League Origins of an American Obsession.* He argues that three Ivy League schools—Princeton, Harvard, and Yale—played foundational roles in the creation of American football. "Princeton had played the first intercollegiate game and initiated both early rules conferences. Harvard had pioneered the rugby style of game that turned American football away from soccer. And Yale quickly became the acknowledged leader in the development of rules and, for two generations thereafter, the sport's flagship program."

In the 1880s, amateur athletic clubs picked up the game from college players. No professional football teams yet existed. The first person paid to play football was William (Pudge) Heffelfinger, who received $500 from the Allegheny Athletic Association on November 12, 1892, to help it defeat the Pittsburgh Athletic Club. The teams tied.[19]

Football became morally tainted in its earliest days, both because of player injuries and because it quickly became a source of income for the colleges where it was played, Bernstein argues.[20]

Holmes Field, ca. 1886–1887. Holmes Field was the site of many Harvard athletic events in the late 1800s. *Source:* Charles N. Cogswell Collection, Cambridge Historical Commission.

The Harvard faculty abolished football in 1884, a decade after that first match with McGill, because of its risks. "The game of foot ball [sic], as then played in intercollegiate games, had become brutal and dangerous, and that it involved not only danger to life and limb, but what was much more serious, danger to the manly spirit and to the disposition for fair play on the part of the contestants," faculty members argued. Professors begrudged the amount of time players spent practicing and playing games, both in the spring and fall.[21] Despite faculty opposition, games resumed a year later.[22]

Football had become so broadly popular that in 1893 Yale played Princeton in Manhattan on Thanksgiving Day before forty thousand devoted fans, an event that became an annual tradition for years.[23]

By the turn of the century, the rowdy football games, played on the northern edge of Harvard's rapidly expanding campus, needed a new home. The university built a stadium across the Charles River in Boston for the considerable sum of $325,000. Harvard Stadium, its design inspired by the Colosseum in Rome, opened in 1903 with the capacity to seat forty thousand. It was "a striking monument to the status achieved by

Harvard football team, 1890. *Source:* Library of Congress.

the sport invented less than thirty years earlier," according to one writer.[24] Harvard President Charles Eliot, no fan of football, tried to get the competition shut down. "It is childish," Eliot wrote in his 1904–1905 annual report, "to suppose that the athletic authorities which have permitted football to become a brutal, cheating, demoralizing game can be trusted to reform it."[25] He was overruled by the Harvard Corporation and Overseers, who believed promises that the game would be improved.

JULIA CHILD: *THE FRENCH CHEF*

America is a nation fascinated—perhaps obsessed—with good food. The Food Network and the Cooking Channel reach hundreds of millions of cable television households with a twenty-four-hour buffet of how-to shows, cooking competitions, and celebrity chefs. Grocery stores offer artisanal breads, olive bars, and chef-prepared takeaway foods. In the 1960s, cookbook author and TV chef Julia Child helped create this culture, saving America from endless nights of uninspired and poorly cooked meals, heavy with processed meats, boiled starches, and canned vegetables.

In 1963, Child, who had recently moved to Cambridge from Paris, began broadcasting her enthusiasm for fine cooking from the studios of WGBH, Boston's public television station. She offered encouragement to home cooks, persuading them that fresh ingredients and herbs made better-tasting meals and that skillfully prepared traditional French food was both within reach of the average homemaker and worth the effort. From her televised kitchen, she spoke with confidence as she introduced America to then unfamiliar dishes like boeuf bourguignon, coq au vin, and the omelet. For Child, food was not just fuel; it offered sensual pleasure and embodied the art of good living.

Though she was made famous by her TV show *The French Chef*, Child was neither French nor a chef. Instead, she was a wonderful teacher, according to Chris Kimball, the Boston-based cooking enthusiast, magazine editor, and host of public television's *Christopher Kimball's Milk Street*.[26] "She was so enthusiastic. She was learning as the audience learned. She went on this voyage of discovery," says Kimball.

Kimball argues that Child, paradoxically, came to prominence at a time when American interest in preparing elaborate meals was subsiding. French cooking requires patience and expertise. By the late 1960s, more American women had begun to explore their talents outside the domestic sphere. "Julia Child didn't make any sense," Kimball says, yet she convinced people like him "that I needed to spend all night Friday and all day Saturday cooking for those dinner parties . . . because Julia said we should." It wasn't

Julia Child at the taping of an episode of *The French Chef* involving the roasting of chickens, April 1970. *Source:* Courtesy of the Schlesinger Library, Radcliffe Institute for Advanced Study, Harvard University.

just the food that brought her an audience. "My conclusion about Julia is, it wasn't the French cooking, it was Julia. Julia could have told you how to use a rototiller."

Julia McWilliams was born in Pasadena, California, in 1912 into a wealthy family. As she recounted in the autobiographical *My Life in France*, her family exhibited no particular interest in fine food or skill in cooking. A young woman of her socioeconomic standing was not expected to develop a passion for cooking, any more than she would be expected to master carpentry or dressmaking. Like her mother before her, Child attended Smith College in Northampton, Massachusetts, graduating in 1934 with a degree in history.

During World War II, she joined the Office of Strategic Services (OSS) in Washington, DC, the forerunner to today's Central Intelligence Agency, working as a typist, file clerk, and research assistant. In 1944, the OSS sent her to Ceylon (modern Sri Lanka) and later China. There she met Paul Child, a fellow OSS employee. The two were married in 1946, she at the age of thirty-four, he a decade older. In 1948, the Childs moved from Washington, DC, to Paris, where Paul worked for the US Information Agency, an arm of the State Department. A skilled artist and photographer, Paul Child was tasked with organizing art exhibits and other cultural events in France. These were intended to strengthen the sense of kinship between the French people and the United States.

In *My Life in France*, Child describes her first meal after arriving in Normandy by ocean liner: a lunch of *sole meunière* that was a transcendent experience. "I closed my eyes and inhaled the rising perfume. Then I lifted a forkful of fish to my mouth, took a bite, and chewed slowly. The flesh of the sole was delicate, with a light but distinct taste of the ocean that blended marvelously with the browned butter. I chewed slowly and swallowed. It was a morsel of perfection."[27]

Inspired by that meal, she took classes at Le Cordon Bleu, the famous cooking school in Paris, earning a diploma in 1951.[28] As her skills improved, she began offering cooking demonstrations and classes to her network of French and American friends in Paris. She encountered fellow Cordon Bleu graduates Simone Beck and Louisette Bertholle, who were attempting to write a book on French cooking for an English-speaking audience. Child joined them, and the three established a small cooking academy called "L'école des trois gourmandes," which Child translated as "school of the three happy eaters."

Their magnum opus, a book entitled *Mastering the Art of French Cooking*, took nearly a decade to complete. *Mastering*, published in 1961, covered the basics of French cooking, with 524 detailed recipes, some quite elaborate. The book was revolutionary in providing accurate measurements of ingredients, precise instructions on technique, and fidelity to authentic French cuisine. Child, her coauthors, and her editors tested the recipes at home to ensure that they were relatively foolproof and could be made with ingredients available in American grocery stores.

The book was an instant hit; within three months it had sold twenty thousand copies. Julia Child's timing couldn't have been better. In January 1961, John F. Kennedy and Jackie Kennedy took residence in the White House, bringing a young, fashionable, modern vibe and an appreciation for French culture to the American mainstream. A second volume of *Mastering* written by Child and Beck was published in 1970. Both volumes remain in print a half century later.

Returning from Europe in 1961, Julia and Paul Child settled in Cambridge, home of Julia's friend and pen pal Avis DeVoto. Julia lived in Cambridge for the next forty years, returning to her native California in 2001, where she lived until her death three years later.

It was the television program that Julia launched (at the age of fifty-one) for Boston's WGBH that made her an American icon. Her television debut was a 1962 guest appearance on a local book review show in which Child promoted *Mastering the Art* and prepared an omelet. Viewers loved her, and she was offered a show of her own. When *The French Chef* debuted in 1963, the concept of commercial-free television was still in its infancy. The first episode was recorded at the demonstration kitchen of a local utility company and ran in February 1963.[29] Later that year, WGBH moved to permanent studios in the Allston neighborhood of Boston, where the remainder of the *French Chef* episodes were recorded. During its ten-year production run, Child recorded more than two hundred half-hour shows.

Early on, the shows were taped without breaks or retakes, so if she made a mistake or dropped a pan, there was no do-over. Julia and Paul prepared meticulously for each show, and she learned to improvise when things went awry, such as the time she dropped a whole chicken on the floor. Her humanity and charm were central to the show's appeal. For the Public Broadcasting System, launched in 1970 as a loose affiliation of the nation's public television stations, Child was a gold mine, bringing in audiences that the network's array of dry educational and public affairs programs did not. More than fifty years after the show's debut, episodes of *The French Chef* remain available on the websites of public television stations nationwide and are regularly rebroadcast.

After *The French Chef* ended in 1973, Child took a break from her public television career but returned with *Julia Child & Company* (1978–1979), *Julia Child & More Company* (1980–1982), *Dinner at Julia's* (1983–1985), and *Cooking with Master Chefs* (1993–1994).

For Child, cooking was all about fundamentals: finding high-quality, fresh ingredients in American supermarkets that were otherwise filled with processed foods;

properly using the right tools—the pots and knives and other implements that she referred to as her *batterie de cuisine*; and employing cooking techniques that would maximize flavor and palatability. At the end of each episode, Child served the week's meal on colorful plates in her television-studio dining room and paired it with wine. Presentation was important, too.

Child's command center was the modest kitchen Paul custom-designed for her in their home on Cambridge's Irving Street. The kitchen became the actual set of three of her television series, including thirty-nine episodes of *In Julia's Kitchen with Master Chefs*, which aired from 1994 to 1996; *Baking with Julia*, thirty-nine episodes of which ran between 1996 and 1999; and twenty-four episodes of *Julia and Jacques Cooking at Home*, which first aired in 1999.[30] Her home kitchen became sufficiently famous that she donated it to the Smithsonian Institution's National Museum of American History in Washington, DC, in 2001, where it remains on display.

Julia Child in her home kitchen on Irving Street in Cambridge. *Source:* Courtesy of the Schlesinger Library, Radcliffe Institute for Advanced Study, Harvard University.

Child, by her example, helped to redefine what it meant to be a smart, ambitious middle-class woman in mid-twentieth-century America. At 6 feet, 2 inches tall, she was not conventionally beautiful or graceful. Her distinctive voice, mocked and immortalized by Dan Aykroyd on *Saturday Night Live*,[31] was not characteristically feminine. Her intelligence and humor were on full display on national television, not masked by false modesty or overshadowed by a male cohost. Child didn't defer to anyone, and she wasn't derailed by dropping a raw chicken in front of the nation. Child was empowered by her mastery of a field—sophisticated French cooking—traditionally dominated by male chefs, and she convinced her viewers that they could be too.

In Cambridge, Julia Child became a prominent local figure. She bought many of her provisions at Savenor's, a butcher shop and market near her home. She carved the words "Bon appetit" into the fresh concrete of a newly laid sidewalk in front of the store; the phrase has since been recast in stone. Her friends and neighbors in the leafy Irving Street area included the economist and Kennedy adviser John Kenneth Galbraith and historian Arthur Schlesinger Jr. She dined at the Harvard Square restaurant Harvest and became friends with its owners, Ben and Jane Thompson, who supplied the cooking implements for her *batterie de cuisine* and the serving platters and tableware she used on *The French Chef*. She mentored, inspired, or collaborated with chefs who shared her love of fine food and carried on her mission, befriending Boston chefs such as Lydia Shire, Barbara Lynch, and Jody Adams. Child also had a fierce competitive streak, Kimball says.[32] She once asked him if he'd actually baked the lemon tart he brought to her house.

She also tested people. One day in the mid-1990s, shortly after he'd moved to the region, Kimball answered the phone and heard that distinctive voice. At first, he thought a friend was pulling a prank, but it really was Julia Child, eager to meet him and inviting him to dinner the following night. When he walked into her kitchen, she thrust a bucket of oysters at him and instructed him to start shucking. This was not one of his skills. After a few fumbling minutes, he declined the task and took solace in a glass of wine. He redeemed himself at another dinner, he says, when she asked him to carve a leg of lamb. "Fortunately, I did know how to do that after failing the oyster test," he says.

Mostly, Kimball says, he was struck by how unpretentious Child was, despite her fame and success. Her number was listed in the phone book, her kitchen had very little counter space, and her own dinner party menus weren't extravagant, he says, remembering one evening of nothing but boiled potatoes, caviar, and wine. "It wasn't like she pulled out all the stops," he says.

But both for himself and for her television audience, Kimball says, "there was this connection with her because she was so authentically Julia."

Julia Child's signature sign-off from her television show—"Bon appetit"—embedded (with her initials) in the sidewalk in front of Savenor's Market on Kirkland Street in Cambridge, where she often shopped, 2019. *Source:* Photo by the authors.

BEN THOMPSON AND THE FESTIVAL MARKETPLACE

The 1970s were a dark time for many American cities. The suburbanization of the United States, which accelerated after World War II, drained older urban centers of middle-class residents as well as employers and retailers, leaving behind impoverished and often racially segregated communities.[33] Cleveland shed a third of its residents between 1960 and 1980. Detroit lost 28 percent, Boston, 19 percent, and Baltimore, 16 percent. By the late 1960s, many urban dwellers grew resistant to the government-funded urban renewal initiatives that had torn apart their cities in the name of saving them. At the same time, federal money for such programs dried up, leaving vacant blocks where vibrant neighborhoods and historic buildings had once stood. Financially overextended and suffering a decline in tax revenues from its once-robust manufacturing sector, New York City nearly declared bankruptcy in 1975 and sought emergency federal aid. "Ford to city: drop dead," read one headline, though President Gerald Ford later denied both the words and the sentiment.[34] Nightly television news programs

Faneuil Hall in Boston, prior to its renovation into a festival marketplace. *Source:* Lois M. Bowen Collection, Cambridge Historical Commission.

portrayed urban areas as places of social disorder, crime, poverty, and corruption. Was there any hope remaining for American cities? Or as a 1971 billboard asked: "Will the last person leaving Seattle turn out the lights."[35]

Where others saw only a crisis, Cambridge architect Benjamin Thompson and Maryland real estate developer James Rouse sensed an opportunity. They combined compelling design and business savvy to create a new type of urban shopping and entertainment venue called the "festival marketplace." Their first collaboration was the Faneuil Hall Marketplace, which opened in Boston in 1976, reusing four historic— but at the time, rather decrepit—market buildings at the edge of downtown. The

Faneuil Hall Marketplace in Boston, 2007. *Source:* Photo by the authors.

reinvigorated market featured stalls selling fresh fish and flavored popcorn, interspersed with handicraft vendors and locally owned restaurants. Jugglers and balloon artists roamed cobblestone-paved pedestrian plazas between the renovated buildings.

Thompson and Rouse merged conspicuous consumption with street theater and a vision of what urban living might ideally be like: densely packed, exciting, and pedestrian-oriented, yet safe, clean, and well-managed. They went out of their way to ensure that the Faneuil Hall project—also known as Quincy Market—felt unique to Boston, not like a suburban shopping mall, in both its materials and retail offerings. The historic character of the city was deliberately placed on display. For some crit-ics, the result was an urban Disneyland, a sanitized and inauthentic consumer-driven experience of Boston, designed to attract out-of-towners while the underlying urban challenges of poverty, disinvestment, and racism remained.[36]

For Thompson, Quincy Market was not a cynical attempt to merchandise the city but the result of a core belief in a European-style pedestrian-scaled human experience, filled with alluring aromas, colorful fabrics, and interesting people, an experience all but lost in car-centric frozen-dinner-eating America.[37] Harvard urban design professor Alex Krieger said, "Ben Thompson was among the first Modernists to figure out the power of intertwining history, commerce, and leisure in the cause of contemporary urbanity." Quincy Market worked on several important levels: it attracted new visitors to Boston, it was a financial success for developer Rouse, and it generated tax revenue for the city.

For cities desperate to attract tourists and retain downtown office workers, the festival marketplace came to be seen as a sure-bet formula for urban revitalization in the 1980s. Reflecting the smaller-government ethos of the Reagan era, these developments relied on a mix of public and private capital rather than tax revenue alone. Local municipalities often footed the bill for infrastructure improvements, trash removal, and policing, while private real estate developers brought investors, marketing expertise, and a design vision.

Following the success of Quincy Market in Boston, Thompson and Rouse collaborated on the creation of Harborplace in Baltimore (opened in 1980), South Street Seaport in New York City (1984), Jacksonville Landing in Florida (1987), Union Station in Washington, DC (1988), and several other urban retail venues. All told, some thirty cities across the United States built festival marketplaces. Five years after its opening, architecture critic Robert Campbell called Quincy Market "an immense boon to Boston, a gift the city hardly deserved considering the fact that its own banks refused to finance a risk as dangerous as the markets were thought to be."[38] Forty years later, Faneuil Hall remained among the most visited tourist destinations in Boston.

Thompson, Quincy Market's design visionary, was something of a serial innovator; the festival marketplace was neither his first nor his only big new idea. In 1945, following service in the US Navy and a degree from the architecture school at Yale University, Thompson cofounded the Architects Collaborative (TAC) in Cambridge. He started the firm with pioneering European modernist architect Walter Gropius and six other partners, all of them in their late twenties.

In 1949, TAC designed a development of modernist single-family houses in nearby Lexington, Massachusetts, called Six Moon Hill, where seven of the eight partners moved their own families. Six Moon Hill was a radical departure from the buttoned-up New England architecture of wooden clapboard houses with gabled roofs and Colonial detailing. To establish overall design coherence, most of the houses at Six Moon Hill had flat roofs, informal plan configurations, and a sense of continuity between indoor and outdoor spaces. The new neighborhood, set in a hilly woodland, embodied the firm's communitarian principles and included four acres of land held in common and a

Ben Thompson at home on Six Moon Hill in Lexington, MA, 1964. *Source:* Photograph by Phillip A. Harrington, used with permission.

shared swimming pool. Thompson designed his own house for maximum flexibility, with partition walls that could be removed as the family's needs changed.[39] Together with his first wife, Mary Okes, he raised five children there. Thompson's son Anthony described the house as "perhaps the worst building Ben Thompson ever designed" but one that he nonetheless loved.[40]

In 1953, Thompson started a retail store called Design Research on Brattle Street in Cambridge's Harvard Square. His goal was to bring well-designed European modern housewares, furniture, and clothing to an American market. The company directly inspired several "lifestyle" retailers, most notably Crate & Barrel and Design Within Reach. The store became known for its practical but stylish modern wares. Design-savvy customers in the Boston area flocked to Design Research. In 1959, it was the first retailer in the United States to sell products from Marimekko, the Finnish fabric and clothing design house. In 1960, First Lady Jackie Kennedy, a paragon of personal style in the post–World War II era, posed for a *Sports Illustrated* cover in a Marimekko dress purchased from Design Research.[41] Though Design Research never became a major chain, the company opened stores in New York City in 1961 and San Francisco in 1965.

The Design Research Building on Brattle Street in Cambridge, 2020. The building was designed by Ben Thompson in the 1960s for his pioneering housewares retail store and his architecture practice. To the right, on Story Street, sits the former headquarters of the Architects Collaborative, the design firm founded by Thompson, Walter Gropius, and six others in the 1940s. *Source:* Photo by the authors.

Thompson practiced what has been called "Cambridge modern" design, heavily influenced by the avant-garde European architecture of the early twentieth century but incorporating regionally derived building practices and materials. "The Cambridge academic and research communities, which drew Gropius here in the first place, believed strongly in the potential of Modernism to physically and intellectually reshape America and the world into the utopia that had always been held out as the ultimate goal of the Total Design project promoted by the Bauhaus," according to architect David Fixler, who has called Gropius "the Pied Piper of Modernism."[42]

Of the architects considered part of the Cambridge modern community, many taught at Harvard's Graduate School of Design. In 1963, Thompson became chair of the school's architecture department, the position formerly held by Walter Gropius. None became household names or what we might call today "starchitects," but designers such as Josep Lluís Sert, Hugh Stubbins, Gropius, and Thompson launched busy Cambridge design practices and built projects across the globe.

Ben Thompson's Design Research retail store in Cambridge, ca. 1964. *Source:* Photograph by Phillip A. Harrington, used with permission.

In 1965, Thompson left Six Moon Hill, moving to Harvard Square. The following year, he left TAC as well, starting his own firm, Benjamin Thompson & Associates. Around the same time, he met Jane Fiske McCullough, who was to become his wife and creative partner for the remainder of his life.

In 1969, Ben and Jane Thompson built a new headquarters for Design Research (D/R) in Harvard Square. The modern concrete structure embodied many of the Thompsons' ideas about design, retailing, and the urban experience that they soon put to use at Quincy Market. The 1969 building was notable for its floor-to-ceiling plate glass windows showcasing the colorful products for sale. An interior design magazine remarked that "Thompson's weather-protected bazaar is a civic space alive with color and visual surprises."[43]

The building housed Thompson's architectural offices as well as the store. Possibly having overextended himself with the construction of the new building, Thompson lost financial control of Design Research in 1970, and the retailer ceased operations in 1978.

Not content to limit their focus to architectural design and retailing, in 1975 the Thompsons opened Harvest, a restaurant in Harvard Square. The restaurant matched their personal style: informal and friendly while also sophisticated and visually appealing. The venue was one of the earliest on the East Coast to feature a farm-to-table dining experience, with the menu changing based on what locally grown produce and locally raised meats were available and fresh. One original menu item—smoked scallops with kiwi raspberry beurre blanc[44]—was radical stuff for the mid-1970s. Alice Waters, the California restaurateur, is widely acknowledged as the inventor of farm-to-table dining in the United States, having opened Chez Panisse in Berkeley in 1971, but the Thompsons weren't far behind. Julia Child, the television chef and Cambridge resident, considered Harvest her favorite restaurant. Many of Boston's best-known chefs and restaurateurs completed tours of duty in the kitchen at Harvest.[45]

In 1992, Ben Thompson received the Gold Medal from the American Institute of Architects, its highest award for an individual architect. In ill health, Thompson retired from his namesake firm in 1993 and spent many of his remaining years in Barnstable on Cape Cod. He died in his Cambridge home in 2002 and was buried at Mount Auburn Cemetery. Quoted in the *New York Times* obituary, architect Moshe Safdie commented that Thompson's career "was an extraordinary celebration of design, life, urbanism and all the things we tend to take for granted now. He was one of the forces that changed America in that respect."[46]

LOVE STORY

Love Story, the 1970 motion picture starring Ali MacGraw and Ryan O'Neal, was nominated for seven Academy Awards and became one of the fifty top-grossing American movies of all time, earning more than $130 million at the box office after production costs of just over $2 million.[47] The American Film Institute has called *Love Story* one of the most romantic movies ever made.[48] In a 2000 article, *Variety* called *Love Story* "the first of the modern-day blockbusters."

The fictional plot of *Love Story* is reasonably straightforward: Jenny Cavilleri, a Radcliffe College student from working-class Cranston, Rhode Island, meets to-the-manor-born Harvard undergraduate Oliver Barrett IV. The two fall madly in love, defying the wishes of their families. Oliver's father disowns him when the young couple marries, withdrawing his Harvard Law School tuition. Jenny gets a job teaching at a local private school to support Oliver, and the two take on the world together. Jenny contracts an unspecified but fatal illness—"movie-wasting disease," as longtime *Boston Globe* movie critic Ty Burr calls it—with Oliver proving that the bonds of romantic love cannot be torn by tragedy. The melodrama begins with the film's opening lines: "What can you say about a 25-year-old girl who died? That she was beautiful. And brilliant. That she loved Mozart and Bach. And the Beatles. And me."

The movie, set in academic Cambridge, catches the colors of late fall, all rusty tweeds, muted ivy greens, grey skies, deep red brick, oak paneling, and dark leather. For many years, Harvard's Crimson Key Society showed *Love Story* each September for incoming freshmen, adding irreverent commentary during the screening.

For the troubled America of the Vietnam era, beset by political strife and the very real deaths of its sons in war, *Love Story* served as a sweetly contrived balm, a way to grieve the country's lost innocence. "It had a cultural impact as a sort of retrogressive cultural movie," says Burr.[49] "It was derided by anybody hip—it's so corny—but they went to see it anyway." Burr remembers being thirteen years old when the film came out and being shocked that so many people loved the movie—including his counter-culture older sister and his more restrained mother. The film's intergenerational popularity proved that as far apart as older and younger generations seemed in 1970 when the Vietnam War was at its height, they could still come together—tissues in hand—in a movie theater. For Hollywood, confused by the unexpected success of *The Graduate* the year before, *Love Story* proved, Burr says "that the old stories still worked."

Audiences adored the story's simplicity and sweetness. Despite knowing what's coming, it's hard not to tear up when Jenny dies and Oliver mourns. Even some high-brow critics gushed. "My admiration for the mechanics of it slops over into a real admiration for the movie itself," wrote Vincent Canby, reviewing *Love Story* for the *New York Times*.[50] Ali MacGraw and Ryan O'Neal are charming to watch. The stirring film score, by French composer Francis Lai, is capable of inciting spontaneous weeping. The story unfolds at a fast clip in only a hundred minutes, leaving little time for viewers to search for deeper meanings.

Love Story's screenplay was written by Erich Segal, who spent a decade at Harvard earning bachelor's, master's, and doctoral degrees and who (rather incongruously for a Hollywood sensation) was a professor of ancient Greek and Latin literature at Yale. After writing the script and selling it to Paramount Pictures, Segal wrote the companion novel, which was released just before Valentine's Day 1970.[51] The book spent more than a year on the *New York Times* best-seller list and sold tens of millions of copies.[52] Canby, the *Times* critic, dismissed it as "almost unreadable."

Several scenes in the movie *Love Story* were filmed in the Radcliffe Quadrangle in Cambridge. The fictional character Jenny lives in Briggs Hall, a Radcliffe College dorm. The book was published on Valentine's Day in 1970 and the film was released on December 16 the same year. *Source:* Photo by the authors.

In some ways, *Love Story* marked Cambridge's cinematic breakout moment, a taste of Hollywood on the Charles River. For the production, Harvard University allowed the film crew liberal access to its grounds. Although the movie opens in New York's Central Park, the Radcliffe Quadrangle off Garden Street has a cameo role as the setting of Jenny's library job and college dormitory. Scenes were shot in Harvard Yard and around Harvard Stadium. A wood-framed house on Oxford Street serves as the fictional newlyweds' new home, where the smitten Oliver carries Jenny across the threshold and up the stairs. Fifty years later, a nearby business still proudly displays a sign in its window: "This laundromat was in the movie *Love Story*."

Cambridge has since starred in or inspired other notable movies, including the following:

- *The Paper Chase* (1973), a tale of academic and romantic challenges set at Harvard Law School, which won actor John Houseman an Oscar for his role as the fictional Professor Kingsfield.

A laundromat on Oxford Street in Cambridge featured in the movie *Love Story* still has a sign in the window fifty years later. *Source:* Photo by the authors.

Actors Ali MacGraw and Ryan O'Neal returned to the Harvard University campus in 2016, nearly five decades after starring in the movie *Love Story*. *Source:* Photo by Jon Chase/Harvard University. Used with permission.

- *Good Will Hunting* (1997), the story of a twenty-year-old South Boston math genius who works as an MIT janitor, written by and starring Cambridge natives Matt Damon and Ben Affleck. The movie won two Oscars, one for Best Supporting Actor (Robin Williams) and another for Best Original Screenplay (Damon and Affleck). *Good Will Hunting* led the relatively unknown actors, childhood friends and fellow graduates of Cambridge Rindge and Latin School, to Hollywood stardom. Burr says it was the first movie that "planted the flag in the larger national consciousness of Boston as a location" for films. "That's when you start getting this wave of new Boston movies. The Boston bro movie by now is a cliche. For my money it starts there," with *Good Will Hunting*.
- *Legally Blonde* (2001), a Reese Witherspoon vehicle set at Harvard Law School (though most of the movie was shot in Los Angeles), which has spawned several sequels and a Broadway show.
- *The Social Network* (2010), written by Aaron Sorkin, lightly fictionalizes Mark Zuckerberg's time as a Harvard undergraduate and the early days of Facebook. The film won three Academy Awards.

After the filming of *Love Story* interfered with Harvard's day-to-day operations, the university limited Hollywood's access to its campus.[53] "I would imagine that the huge success of *Love Story* only tightened the resolve to not be involved" in major motion pictures, Burr says. "It was a big hit, and it was so corny. It's not the kind of thing that Harvard would want to be associated with. If it had been a movie with at least pretensions to honoring the life of the mind . . . but it was Ali MacGraw dying."

The *Love Story* actors, MacGraw and O'Neal, returned to the Harvard campus in 2016 to celebrate the forty-fifth anniversary of the film's release. The two arrived for an event at Harvard's Kirkland House in an antique MG convertible, similar to one featured in the movie, and spoke to a roomful of undergraduates.[54] According to an Associated Press report, both MacGraw and O'Neal admitted they had a crush on each other during filming, though both were then married. "Ryan and I clicked immediately," said MacGraw. "We just had a chemistry."[55]

But not the kind of chemistry usually taught at Harvard.

T. BERRY BRAZELTON AND BEHAVIORAL PEDIATRICS

In the late twentieth century, pediatrician T. Berry Brazelton influenced the way many Americans raised their children. Brazelton had a remarkably productive career, including writing or cowriting more than thirty books on child care; hosting a cable television series for more than a decade; teaching at Harvard Medical School and Boston Children's Hospital; and running a pediatric practice in Cambridge, where he treated more than 25,000 patients over more than half a century.

He taught parents how to observe, listen to, and learn from their babies—and to understand that those parent-child interactions were crucial to a healthy relationship. In 2007, he told a National Public Radio interviewer: "To me, that's the biggest gift I can give to each parent: Watch your baby and trust that baby to tell you when you're on the right track and when you're not."[56]

Brazelton was, at root, an advocate for the emotional well-being of children and their parents. "He always saw the good in people—the strength in the mother, the strength in the babies. He assured me it's there in everybody, otherwise, they don't live," says Heidelise Als, a newborn psychologist who trained under Brazelton at Boston Children's Hospital.[57]

For the field of pediatrics, which traditionally had focused on fighting illness rather than studying children's behavioral development, Brazelton's approach was novel. His books and television programs on child rearing, particularly in the 1980s and 1990s, came at a time when the nature of parenting and parental roles were themselves being redefined. From the early 1970s onward, opportunities for American women to work outside the home were expanding, while many of the classic childhood diseases—measles, mumps, whooping cough, diphtheria, and scarlet fever—were largely controlled through vaccination and antibiotics. Brazelton evolved with the times on issues such as child discipline, two-income families, parental leave policies, high-quality daycare, the importance of play, and the involvement of fathers in child rearing.

Commenting on his *New York Times* obituary, one mother said that reading Brazelton's books helped her learn how to raise children differently from the way she'd been parented.[58] "Dr. Brazelton broke the cycle of violence in my family," wrote the woman from Akron, Ohio. "I never hit my children and learned from him that if I

Pediatrician T. Berry Brazelton. *Source:* Courtesy, Brazelton Touchpoints Center, Children's Hospital Boston.

were an adult, I could think of more effective non-violent ways of raising my children. The payoff is to see my grown well-adjusted adult children using the same effective, loving techniques I used to raise their children. Thank you, Dr. Brazelton!"

Brazelton was also a leading pediatric researcher, responsible for developing a scale used to study newborn behavior. "As a true scientist, he questioned the reigning scientific paradigm and was very tuned into where its limitations were beginning to be exposed," says Joshua Sparrow, who collaborated with Brazelton for twenty-seven years and took over the Brazelton Touchpoints Center at Boston Children's Hospital after his mentor's death.[59] The newborn scale encouraged doctors and parents to notice that each baby's behavioral pattern was unique and that that behavior served as a form of communication with his or her parents. "He demonstrated that newborns shape their caregiver's behavior as much or more than caregivers shape their newborn's behavior," an idea that was revolutionary for its time, Sparrow says. By putting these behaviors on a formal scale, Brazelton encouraged doctors and parents to perceive babies differently. "It changes you because you can't help seeing things that you never saw before," Sparrow says. When Brazelton began his work, parents with children or babies in the hospital were allowed to visit for an hour a week. Today, they essentially have twenty-four-hour-a-day access and are encouraged to spend as much time as possible. "That change really was a result of his work," Sparrow says.

Brazelton's own childhood in Waco, Texas, where he was born in 1918, likely shaped his views on the relationship between children and their caregivers. His parents were not particularly affectionate, he reported in his 2013 memoir, *Learning to Listen: A Life Caring for Children*. But he became skilled at observing and working with young people. "At every family event, and there were many, I was put in charge of all nine first cousins while aunts and uncles and grandparents prepared for the big dinner."[60] Brazelton also had to look after the family's ducks and chickens. "I think as a child, he did a lot of observing of younger children's and animal behavior," Sparrow says, and he developed unique observational and communication skills.[61] "That is part of what allowed him to see what newborns could do and absorb it and make meaning out of it," Sparrow says. "It also allowed him to be able to tell the world about it and to get the world excited about it."

After completing medical school in New York City and serving in the US Navy during World War II, Brazelton finished his medical residency at Massachusetts General Hospital in 1945 and went through pediatric training at what was then called Children's Hospital Boston. He began practicing as a pediatrician in Cambridge in 1950 and teaching at Harvard Medical School in 1953. In his pediatric practice, he was able to observe the interactions between babies and their parents. He argued that newborns

Former home and pediatric practice of T. Berry Brazelton on Hawthorn Street in Cambridge.
Source: Photo by the authors.

had innate personalities that developed before birth and that these varied from child to child. Babies are not, as was often thought, simply "blank slates." Brazelton noted that each new skill a young child learned—sitting up, walking, talking, using the toilet—could be accompanied by regression in other abilities or behaviors because the effort to learn a new skill is so taxing on a child's energies. He labeled these moments of frustration and growth *touchpoints*.

In the 1970s, he began a research center on pediatric behavior at Boston Children's, where he taught his students how to observe infant-parent interactions. Heidelise Als, the neonatology specialist, was a PhD student at the University of Pennsylvania using Brazelton's observational techniques, when he invited her to a planning meeting for the center.[62] A year later, he hired Als to join his research center, where she trained many of his fellows in the care of premature babies. Brazelton, she says, was the first medical doctor who focused on the field of childhood development, which had mainly included psychologists like herself. "The field needed someone who wasn't just measuring things to death, but seeing the bigger picture," she says, and he played that role.

T. Berry Brazelton holding a small child. *Source:* Courtesy,
Brazelton Touchpoints Center, Children's Hospital Boston.

His laboratory videotaped young children and their parents as a way to study their
interactions. Together with colleague Ed Tronick, Brazelton used frame-by-frame videos to show how a baby—even a two-month-old—reacted when its mother turned
away. What he saw, Sparrow says, was that very early on, infants were active partic-
ipants in their communications with their caregivers. Eight-week-olds can converse
long before they have words. Babies will pick up the parents' rhythm, tone, pitch, pace,
and conversational structure, Sparrow says, even initiating or reshaping the communication themselves. Brazelton took his videos to Congress to lobby for the Family and
Medical Leave Act of 1993. "He was able to change the hearts and minds of people in
Congress," Sparrow says, when they were able to see that even at eight weeks, a baby is
distressed when a parent turns away and is unresponsive.

Brazelton had a ready smile, enormous charisma, and a soothing Texas drawl that enchanted everyone from television viewers to young patients. "There was this unique way he had of being with [a person]," Sparrow says. He made you feel like it was "just you and him on top of the highest mountaintop." Brazelton was a popularizer, which earned him critics in the pediatrics establishment, but he grounded his advice in scientific research. He published more than two hundred scientific papers and book chapters, and he traveled the world to study the ways children were reared in other societies.

For sixty-seven years, Brazelton and his wife Christina lived on Hawthorn Street just outside Harvard Square. From his base in Cambridge, Brazelton raised three daughters and one son, trained hundreds of medical professionals, and cared for tens of thousands of patients. He was as beloved by his own patients as he was by people who read his books or watched him on TV. As Sparrow says, "People who had never met him felt like he had raised their children with them."

CAR TALK

In 1977, Cambridge car mechanics Tom and Ray Magliozzi were asked to join an on-air panel on Boston radio station WBUR to discuss car repair. Ray thought it was dumb idea and declined. Tom agreed to participate and turned out to be a panel of one, as no other mechanic showed up. Pleased with the appearance, the station invited him back. This time, Tom brought his younger brother with him, the beginning of more than three decades of broadcasting together.[63] Originally hosting the show every week for free, the pair eventually asked WBUR for $25 a week. As *Car Talk* gained a local following, they secured the rights to the name and broadcast content.

On their radio program, the Magliozzi brothers—who sometimes referred to themselves as "Click and Clack, the Tappet Brothers"—attempted to diagnose callers' automotive problems and estimate repair costs. Often, the issues crossed into the personal, as in "My husband doesn't like the way I drive his car, but I think he is crazy." In 2000, Tom Magliozzi told the *Boston Globe*: "Somewhere along the line we decided that cars were boring. But they provide an entree into life, so we can talk about life philosophy. Which is more interesting than talking about valve clearances."[64]

In 1987, National Public Radio host Susan Stamberg invited the pair to host a weekly *Car Talk* segment on Sunday—their first exposure to a national radio audience.[65] Later that year, a full hour of *Car Talk* was distributed to NPR stations every week. New episodes of *Car Talk* ran on public radio stations weekly until 2012. Reruns of the program continued to be broadcast and podcast as *The Best of Car Talk*. In 2014,

Brothers Tom (right) and Ray Magliozzi of the *Car Talk* radio show. *Source:* Richard Howard, used with permission.

WBUR reported that *Car Talk* was broadcast on 660 radio stations across the United States, reaching an audience of three million people.[66]

Doug Berman, who became the show's executive producer, says he first realized the program was something special in 1988, about six months after it began airing nationally, when the brothers joined a panel at an annual NPR conference. At the time, most public radio stations were focused on news programming and classical music. "Jazz was about as much as they let their hair down," Berman says. "The people in charge of the stations were trying to figure out if this [show] was something they could live with." Listening to Tom and Ray, the packed audience overflowed with laughter. "It was so good and so infectious that those questions became moot," Berman says.

Ray Magliozzi, Doug Berman, and Tom Magliozzi in the recording studios of radio station WBUR preparing for a broadcast. *Source:* Richard Howard, used with permission.

Car Talk became a cultural phenomenon, spawning a newspaper column, an animated television series (though it produced only ten episodes in its lone season, 2008), and a merchandising arm—the "Shameless Commerce Division," as the brothers regularly called it on air. For local NPR affiliate stations, *Car Talk* served as a reliable generator of listener pledges, the money that supports the advertiser-free radio network. In 1992, *Car Talk* received a Peabody Award, the broadcast industry's most prestigious award for news and nonfiction storytelling.[67] In 2018, the Magliozzis were inducted into the Automotive Hall of Fame, a museum in Dearborn, Michigan, that celebrates the lives of auto industry pioneers.[68]

Many of the show's funniest remarks were comments on one of the brothers' own personal foibles, such as Tom's ill-fated relationships with women and cars. Often, a caller would be asked to simulate on-air the noises his or her car was making, with humorous results. Each week's broadcast ended with the admonition "You've wasted another perfectly good hour listening to *Car Talk*" and the directive "Don't drive like my brother." Tom's raucous and sometimes uncontrolled laughter took up precious airtime, but the pair's lighthearted manner was a key component of the show's charm.

The show was a hit with children, too, even ones too little to know much about cars. "Kids intuitively can tell when adults are misbehaving," says Berman.[69] Children knew that Tom and Ray "weren't supposed to do that on the radio and delighted in the guys' misbehavior."

The Magliozzi brothers were raised in East Cambridge. Their father delivered home heating oil. "They considered themselves sons of Cambridge," says Berman. The brothers were both educated at MIT. In the early 1970s, they opened a do-it-yourself car repair shop in Cambridge called Hacker's Haven.[70] Later, they converted the business to a conventional repair shop, the Good News Garage.

By way of explaining why he traded a promising career in engineering for the car repair business, Tom often told the story of his near-death experience one day commuting to an engineering job in the Boston suburbs. In Berman's memory, Tom was driving a tiny, convertible Triumph TR6 down Route 128, when his right rear wheel came careening off and the axle hit the pavement in a "blaze of sparks." Tom managed to maneuver the swerving car into the entrance of a highway service station and bring it to a stop, as alarmed attendants came running over to help. When they arrived to see if he was all right, Tom said, "Fill 'er up, and check the oil!"

Tom realized he didn't want to die having spent his time "living a life of quiet desperation," as the *New York Times* quoted him in his 2014 obituary.[71] "So I pulled up into the parking lot, walked to my boss's office and quit on the spot." In one retelling, his brother chimed in, "Most people would have bought a bigger car."

On the air, the brothers Magliozzi, speaking in their distinctive Cambridge accents, frequently referred to their hometown as "Cambridge, our fair city, MA." Their production company was located in Harvard Square, and Cambridge itself was often the butt of their jokes.

Berman remembers the day they painted the now-iconic "Dewey, Cheetham & Howe" sign on the window of their third-floor offices. The trio chose Dewey, Cheetham & Howe as the name of their production company because they thought it sounded like a law firm and would be a funny way to answer the phone. They had just moved in and had nothing in the office but a table and a few folding chairs. Berman hired a sign painter to add gold-leaf letters to the window facing Harvard Square, in the style of a law firm, to "complete the joke."

The sign painter, an elderly man, shuffled into the nearly empty office, stooped over, carrying his toolbox. He borrowed a folding chair to stand on, as Berman remembers it, and set to work. The Magliozzi brothers and Berman got busy discussing the newspaper column they were writing that week. A reader had asked a question about car insurance, and Tom and Ray went on an extended riff about how awful the insurance

The window of the *Car Talk* production offices in Harvard Square, jokingly labeled as the fictional law offices of Dewey, Cheetham & Howe, 2015. *Source:* Photo by the authors.

industry—and everyone in it—was. They were all laughing hysterically by the end, Berman says. There was a pause as they caught their breaths. The sign painter, who they'd forgotten was there, then gingerly stepped down off his chair, looked at them, and said, "You know, my brother was in the insurance industry." The three fell silent, convinced that they had terribly insulted the man they were relying on to make their sign. "He screwed everyone!" the man said, using a different word for "screwed." The laughter came even louder than before. "It was a great moment," Berman says. And the sign painter must have done a good job—"because it lasted all these years."

The City of Cambridge's Historical Commission secured an agreement with the company that redeveloped the building in 2019 to keep the sign in place in perpetuity, Berman says. "So if you ever see it go away, get in touch with me. I have the paperwork!"

After a lengthy battle with Alzheimer's disease, Tom died in 2014 at the age of seventy-seven. Ray and Berman have kept the show alive with reruns and short updates, and as of 2020 Ray continued to write the weekly newspaper column.

Asked what he thinks of as the essence of the Magliozzi brothers, Berman talks about a thank-you letter the show received from a listener. The author's father had

Ray (left) and Tom Magliozzi of *Car Talk* in the Good News Garage, the car repair shop in Cambridge run by Ray. *Source:* Richard Howard, used with permission.

Tom and Ray Magliozzi of *Car Talk* delivering the keynote speech at MIT's graduation ceremonies, 1999. *Source:* Richard Howard, used with permission.

been lost to the fog of Alzheimer's several years earlier. She was driving him to an appointment with *Car Talk* playing on the radio when he started laughing for the first time in forever. She realized he was laughing with the *Car Talk* guys, and as she looked at him, she started laughing too. They laughed and connected in a way they hadn't in years. And as they were looking into each other's eyes and laughing, she rear-ended a car that had stopped car in front of her. "But it was worth it!" she said in the letter.

"I think that defines them well," Berman says of Tom and Ray. "They did a lot to bring people together and make people laugh and get away from the troubles of their lives, whether they were car troubles or any troubles. It's why a lot of people still like listening to the programs today."

YO-YO MA: CONNECTING CENTERS AND EDGES

On a cold night in early 2020, fans crowded a Harvard Square theater to celebrate the twentieth anniversary of Silkroad, a Boston-based project that brings together musicians from cultural traditions across the globe. Performers included a Chinese American violinist, an Indian drummer, a Galician bagpiper, and a man playing a Japanese wind instrument called a shakuhachi. The audience's enthusiasm was palpable, particularly for two groups of visiting Native American high school students who joined the evening's performance. Silkroad's founder, world-renowned cellist Yo-Yo Ma, sat mostly in a back corner of the stage, smiling broadly and applauding enthusiastically for the students and his musical colleagues. In an age of outsized celebrity egos, it was almost surprising to see Ma cheering others at center stage, content to remain on the margins.

Though he regularly travels the world as a musician, Ma has lived much of his adult life in Cambridge, where the kind of cultural diversity on display at the Silkroad performance that winter night is not considered unusual.

In his recordings and performances, Ma has explored the musical traditions of Argentina, Brazil, the Appalachian region of the United States, and the lands that in ancient times comprised the Silk Road trade routes stretching from China to Europe. At the same time, Ma is lauded as a skilled interpreter of European classical music, especially the work of Johann Sebastian Bach, whose cello suites he has recorded three times. In the course of a career that began with his first public performance at the age of five, Ma has recorded more than ninety albums, received eighteen Grammy awards, and become arguably the best-known classical musician working today. In 2006, the United Nations named Ma a Messenger of Peace in recognition of his ongoing work to bridge cultural divides through musical exchange.

Both of Ma's parents were musicians, his father a violinist and his mother a singer. Born in China, they were living in Paris at the time of Ma's 1955 birth, struggling to make a living teaching music.[72] Ma discovered the cello very early: "When I was four, we went to the Paris Conservatory to visit somebody, and there was an oversized double bass. Just like kids all want to play with fire engines, I saw this double bass—which went up literally as high as the ceiling—and I thought 'This is a really cool instrument.' So, 'Please, please, can I play double bass?' The compromise was the next-to-largest instrument, the cello." Ma began taking lessons with his father and very quickly became a prodigy.

In 1962, the family moved to New York City, where his father was offered a position teaching music at a private school. In lectures and interviews, Ma has embraced his identity as an immigrant, recalling his move from France: "In the space of one day languages changed, values changed, spaces, colours, weather cycles, architecture, person-to-person relationships—they all changed. The world as I experienced it was full of awe, wonder and confusion."[73]

A few years after arriving in the United States, Ma was invited to perform in Washington, DC, for an audience that included former President Dwight D. Eisenhower

Yo-Yo Ma and Kojiro Umezaki of Silkroad after a performance at the American Repertory Theater's Oberon stage in Harvard Square, Cambridge, February 2020. *Source:* Photo by the authors.

and his successor John F. Kennedy.[74] In the televised concert, the seven-year-old Yo-Yo Ma and his eleven-year-old sister Yeou-Cheng Ma were introduced by the composer and conductor Leonard Bernstein. Ma has since performed for President George W. Bush, from whom he received the National Medal of Arts in 2002, and at the inauguration of Barack Obama, who awarded him the Presidential Medal of Freedom in 2011.

Ma began studying at the Juilliard School in New York City at age eight. At sixteen, he enrolled at Harvard University as an undergraduate. Harvard's music department did not offer a conservatory program for training professional musicians; it focused on the theory and history of music and composition. To complement his musical education, Ma took up anthropology, the study of human societies. He told *Harvard Magazine* in 2000, "Harvard was the first place in my life where I was systematically introduced to different worlds and ways of thinking. I learned there how science and art are joined under philosophy."[75] For the first time in his life, Ma noted, he was surrounded by people who were not musicians. In 2009, he told the student newspaper the *Harvard Crimson*, "I always thought that I should do something other than music. . . . Music is part of my life, but there's a life separate from my life as a musician, and that actually is more of my identity. It wasn't until five years ago that I realized that my real passion is people."[76]

By the late 1980s, Ma's recordings began to include artists and works outside the classical canon. In 1989, he recorded an album of Cole Porter songs performed with French jazz violinist Stephane Grappelli. In 1992, Ma recorded an album with jazz vocalist Bobby McFerrin, and in the mid-1990s, he began to explore the music of Appalachia with bassist Edgar Meyer and fiddler Mark O'Connor.

There aren't that many classical works written for solo cello, and a cellist who begins mastering them at age five will soon run out of repertoire. Composer Richard Danielpour, speaking with *Harvard Magazine* about Ma, said, "He's been there for so long—he mastered the instrument so early and conquered every frontier for a cellist. Great artists, especially artists like Yo-Yo with his deeply inquisitive mind and concern for human beings, know they have to keep moving, searching, exploring, so they don't atrophy."[77]

The Silk Road Project, which Ma founded in 1998, is perhaps the most sustained result of Ma's cross-cultural interests. The project brings established musicians from different parts of the world—primarily Eurasia—together to perform, sharing their musical traditions with each other and with audiences. Now known simply as Silkroad, the initiative has commissioned more than seventy new pieces of chamber music and enlists a changing roster of more than fifty musicians in performances around the world. As of 2019, Silkroad had recorded seven albums, one of which, *Sing Me Home*, won the 2016 Grammy for Best World Music Album.[78]

Yo-Yo Ma performing at Harvard, 1974. *Source:* Courtesy of E. B. Boatner and the Schlesinger Library, Radcliffe Institute for Advanced Study, Harvard University.

A billboard in Harvard Yard announcing a Yo-Yo Ma performance, ca. 1970s. *Source:* Courtesy of E. B Boatner and the Schlesinger Library, Radcliffe Institute for Advanced Study, Harvard University.

Ma is noted for his personal warmth and willingness to share, both on stage and off. *New Yorker* music critic Alex Ross noted: "Ma's unfailing generosity of spirit—everything you have heard about the niceness of the man is true—gives substance to his homilies."[79] As a United Nations peace representative, Ma has advocated for treating migrants with dignity.

In a lecture at the Massachusetts Institute of Technology in 2018, Ma shared his thoughts about artistic productivity and the role he believes it can play in effecting social progress. One idea is what he terms "edge-center oscillation"—the need to connect new ideas forming at the fringes of a creative discipline with its mainstream.[80] "Culture helps the edges of society communicate with the center of society," he said.

In performances like the one in Harvard Square that winter night, Ma has helped bring the artistic traditions of marginalized or non-Western cultures to wider attention, even as he remains a master of classical European music. In 2018, Ma brought his career full circle with a new recording and thirty-six live performances over two years of Bach's Cello Suites, pieces he began to learn when he was smaller than a full-sized cello.

RE: INVENTION

How has Cambridge managed to reinvent itself when so many other postindustrial cities in the northeastern United States have not? And how has this city, which has never had more than 125,000 residents, managed to have such an outsized influence on the rest of the country? Researching and writing this book have led us to a few thoughts about the ingredients in Cambridge's "secret sauce," which we outline below, along with cautions for the future.

LESSON 1: EDUCATION BENEFITS NOT JUST INDIVIDUALS BUT THE COMMUNITY AS A WHOLE. THIS IS ESPECIALLY TRUE IN A KNOWLEDGE-BASED ECONOMY. Harvard economist Edward Glaeser argues that the percentage of a city's population holding a college degree is a good predictor of its economic resilience because education and skills prepare people to adapt to change.[1] In Cambridge, more than 72 percent of adults hold a bachelor's degree or higher, compared with 28 percent of Americans overall.[2] "Like skilled people, skilled cities also seem to be better at reinventing themselves during volatile times," according to Glaeser's 2011 book, *Triumph of the City*. When, for instance, some of his peers predicted Cambridge's demise after the decline of its candy industry, Glaeser says that "they underestimated the ability of skilled cities to reinvent themselves."[3]

Cambridge's commitment to education goes back to its founding. The Puritans who first settled Massachusetts believed "that reading the Bible was the surest means of knowing God's will. They saw education as a key brick in that 'Bulwark against the kingdom of AntiChrist.'"[4] In 1636, they invested more than half of the colony's tax revenues to start what became Harvard College. Harvard and MIT remain Cambridge's top two employers.[5]

Healthcare and education helped resuscitate the region's economy after its industrial base disappeared, according to political economist Barry Bluestone of Northeastern University.[6] Between 1950 and 1980, Boston's population shrank from 800,000 to 562,000, and surrounding communities like Cambridge experienced similar declines. But today, that trend has reversed, and Boston's population is closing in on 700,000, thanks to the rise of higher education and healthcare. "Those two industries really led the recovery of Boston/Cambridge," Bluestone says.

The development of new industries used to be credited to lone inventors like Eli Whitney and Alexander Graham Bell, but today, Bluestone says, innovations are more likely to emerge from large groups of researchers. "Most of the major new industries are coming out of universities, and most of the great medical advances are coming out of the great teaching hospitals," he says. "Boston/Cambridge is number one in the nation in both of those." And biotechnology, "which is what's leading Cambridge to where it is today," he notes, is a combination of both.

LESSON 2: CAMBRIDGE'S URBAN DENSITY AND DIVERSITY ALLOW IDEAS TO FLOW IN MULTIPLE DIRECTIONS AND FROM UNEXPECTED PLACES.   Cambridge remains one of the most densely populated cities in America—a few slots below San Francisco but with more people per square mile than neighboring Boston.[7] It is big enough to be a "real" city but small enough that people know their neighbors. Zipcar cofounder Robin Chase says the city's size helped make it a perfect testing ground for her company. "Cambridge is the minimum viable city. It really is a city, and it can do all those things that cities can do," she says. "But it's not Brooklyn or Manhattan. It's got a size that is helpful. It's a friendly and relevant size."[8]

Some research suggests that face-to-face interactions foster innovation in fields focused on information, like New York's finance industry.[9] For information-centric industries like biotechnology, Bluestone says, "proximity and density turn out to be really critical for innovation."

The connection between urban density and innovation is not a new idea. Economic historians like Paolo Malanima have linked rising population density in northern Italy after the Black Death of the 1300s with the development of the Italian Renaissance in art, literature, science, and industry.[10] Italy's population density was higher than the rest of Europe's from the second half of the thirteenth to the seventeenth centuries, according to Malanima: "The impact of human interaction is likely to have been much stronger within this European region than elsewhere in the continent."

Population diversity also fosters innovation, research shows—and Cambridge is far more diverse than the typical American city of its size. In the nineteenth century, the city's population exploded from just 2,400 to more than 91,000, largely driven by immigration from Ireland, Italy, Portugal, and French Canada. In more recent decades, significant numbers of immigrants have arrived from Haiti, the Bahamas, the Dominican Republic, Somalia, Ethiopia, India, China, and other Central American, African, and Asian countries. Just over a quarter of the city's population in 2009 was born outside the United States[11]—compared to a national rate of 13.7 percent foreign-born.[12]

Students at Cambridge's public high school report speaking more than twenty different languages at home.[13]

Bluestone says colleges and universities have helped the Boston area attract particularly productive immigrants. "The people we attract are the best and the brightest from all those countries," he says. "There's an enormous amount of human capital coming here and until very recently, staying here."

Richard Florida of the University of Toronto argues in his 2002 book, *The Rise of the Creative Class*, that the most economically successful cities excel at "technology, talent, and tolerance."[14] "They had clusters of technology industry; they had great school systems and research universities that produced talent; and they were open-minded and tolerant, which allowed them to attract and retain talent regardless of gender, race, ethnicity, and sexual orientation."

LESSON 3: FOLLOW THE MONEY. CAMBRIDGE'S SUCCESS AS AN INNOVATION HUB HAS BEEN FOSTERED BY TAXPAYERS, VENTURE CAPITALISTS, AND PHILANTHROPISTS.   In 2017, according to Barry Bluestone, the Boston-Cambridge region received $2.5 billion in funding from the National Institutes of Health—the most of any region in the country. New York, which has roughly three times the population of metropolitan Boston, received $2 billion, and San Francisco, $1.4 billion. These infusions support the salaries of professors, postdoctoral researchers, lab technicians, security guards, and custodians and indirectly pay for child care, restaurant meals, home improvements, and car repairs.

Both Harvard and MIT have benefited tremendously from private philanthropy, particularly in the area of science and engineering. Beginning in the 1760s, Harvard began assembling a collection of mostly donated artworks, biological specimens, scientific instruments, and cultural artifacts, collectively called the Philosophy Chamber, that became an important early scientific resource.[15] The textile magnate Abbott Lawrence's gift of $50,000 in 1847 established the Lawrence Scientific School at Harvard. At the time, it was the largest single gift ever given by an individual to an American college or university.[16] An industrialist named Gordon McKay, who made his fortune from a machine he licensed to shoe makers, left $4 million to Harvard in 1893 for the purpose of supporting research in the "useful arts"—what we would call engineering.[17]

Over his lifetime, photography pioneer George Eastman, founder of Kodak, donated more than $22 million to MIT—an enormous sum at the time—even though the Rochester, New York, businessman had no obvious connection to the Institute. His initial gift of $2.5 million in 1912 came at a crucial moment when MIT, struggling financially, was considering a merger with Harvard. Eastman's contributions helped to turn MIT from a "local commuter facility to international leader," one author wrote.[18]

Private venture capital has also been a huge factor in the region's success. In 2018, the Boston area overtook New York City to become the second-largest destination (after Silicon Valley) for venture capital investments.[19] In an interview, entrepreneur Robin Chase recounted a story from the early days of Zipcar. She bumped into a friend at a cocktail party, who asked how her startup was faring. She told him that she needed $25,000 in seed capital within twenty-four hours or the business would fail. By 9 the next morning, he'd sent her a check, and she was able to proceed with Zipcar's launch.

LESSON 4: MONEY ISN'T EVERYTHING. SERVING THE GREATER GOOD MATTERS, TOO. Although funding is essential for advancing good ideas—as Elias Howe found out when struggling to make and promote his sewing machine—having a social purpose matters, too. Many of the people we feature fought for something bigger than themselves: racial justice, gender equality, the freedom to marry. Anne Bradstreet was serving God with her poetry. Benjamin Waterhouse sought to save America from the horrors of smallpox. T. Berry Brazelton used his research to improve parenting skills. Robin Chase and Antje Danielson founded Zipcar in part to address climate change. During World War II, Vannevar Bush, Louis Fieser, and Doc Edgerton were motivated to defeat totalitarianism and preserve democratic government. In promoting peaceful uses of biology, Matt Meselson made the world safer for all of us. Noam Chomsky has been the conscience of several generations. Robert Langer transformed the delivery of medicines while trying to do something meaningful with his life. Again and again, we have seen local people pursue passions for reasons other than personal fame and financial success—though those sometimes came, too, the result of doing well by doing good.

LESSON 5: BUILD ON LOCAL STRENGTHS. Cambridge lacks a deep-water port like Boston's. Nor does it have rich soils suitable for agriculture. In 1636, Puritan minister Thomas Hooker and a group of his followers abandoned Cambridge to found a new city—Hartford—along the banks of the Connecticut River, partly in search of fertile soils for farming. When early colonial leaders decided to locate their new college here, it was something of a consolation prize after the colony's capital was permanently moved to Boston. In the 1800s, the city's one inland water body, Fresh Pond, was exploited for its ice. A brick-making industry grew out of the city's extensive clay deposits, and a diversified manufacturing base grew on the cheap land reclaimed from the Charles River basin. In more recent decades, sturdily constructed factories originally built for candy making have been repurposed for pharmaceutical research and high-tech startups. Over the centuries, what were once considered local weaknesses became local strengths, repurposed to new needs.

In a 2013 report, researchers with a Cambridge-based think tank made a list of assets that could be used to spark economic growth in older American cities. The list included traditional, walkable downtown areas; architecturally and historically distinctive neighborhoods; manufacturing; and locally based corporations and business communities.[20] Cambridge in 2020 was able to tick nearly every box. These are assets that can help a community endure a weak economy or shifts in the priorities of government leaders.

LESSON 6: A DIVERSE ECONOMIC BASE HAS HELPED CAMBRIDGE ESCAPE THE SHARP DECLINES OF CITIES THAT DEPEND ON A SINGLE INDUSTRY OR EMPLOYER. Cambridge has long had a diversified economy. The city's industrial base, which began to grow in the early nineteenth century, included an array of regionally and nationally important companies including glassmakers, cookie bakers, book printers, and soap manufacturers. Today, the city's biggest employers include universities, pharmaceutical companies, and digital technology firms.

History shows that relying on a single business can leave a community vulnerable. Over time, big Cambridge employers such as Polaroid and Lotus Development, which once dominated their industries, lost their competitive advantage and shed local workers. Others relocated out of the city, like the Lever Brothers soap company, an important local employer through the 1950s. Competitive pressure in a capitalist society is ruthless, and investment capital is mobile. Although most industrially based cities in the Northeast and Midwest suffered from deindustrialization after World War II, cities with a diversified economic base suffered less.[21] Youngstown, Ohio, for example, was crippled by the loss of its dominant steelmaking industry in the 1970s and still struggles today. Detroit has suffered through every downturn in the car-making industry. Through foresight or just plain luck, Cambridge has managed to host pioneering businesses—Polaroid in the 1930s, Lotus Development in the 1980s, Genzyme in the 1980s and 1990s, Education First (EF) and Akamai in the early 2000s—that became major employers when other industries declined. To remain robust, cities like Cambridge need to continue to nurture small entrepreneurs and a broad economic base.

LESSON 7: IT HELPS TO BE NEXT DOOR TO A BIG CITY LIKE BOSTON. There's no question that Boston has played a major role in Cambridge's continued success. The two cities are connected by nine bridges across the Charles River,[22] and they are intertwined economically, culturally, and socially. Cambridge innovators have benefited from Boston's large customer base, its law firms, its accountants, and its banks. When sewing machine inventor Elias Howe needed an attorney to pursue Isaac Singer for patent infringement, he was able to find a good one in Boston. When biotech startups need money to get off the ground, they can turn to local bankers and venture capital firms, some just across the river. Even the proximity of Boston's Logan International Airport makes a difference. Its direct flights to major destinations around the globe funnel visitors to the region and foster the exchange of ideas and goods—serving the same purpose a deep-water harbor did in the 1850s.

But Boston is not the explanation for all of Cambridge's success. If proximity to a big city were all a smaller city needed to thrive, then Camden, New Jersey, just across the Delaware River from Philadelphia would be a vibrant community instead of one of the country's most hard-pressed. The same would be true for Gary, Indiana, a forty-five-minute drive from Chicago. A small city needs more than proximity to a bigger one to flourish.

LESSON 8: VISIONARY LEADERSHIP MATTERS. The "Massachusetts Miracle" that saved the Boston region from its downward spiral in the 1990s was a combination of the "pure dumb luck" of already having renowned universities and hospitals, "plus some good mayors and some good civic engagement," says Barry Bluestone of Northeastern.[23]

Visionary leadership has repeatedly made a difference in Cambridge's success. MIT's Vannevar Bush and Harvard's James Conant enabled the two institutions to play important roles in the nation's defense during World War II and positioned them for continued federal funding afterward. Cambridge Mayor Al Vellucci, by spearheading a moratorium on recombinant DNA research in 1976, helped lay the groundwork for the later biotech boom. Cambridge City Manager Robert Healy, who led the government from 1981 to 2013, smoothed over town-gown tensions, supported new development, and ably managed the city's fiscal resources. On the business side, the decision in the early 2000s by drug giant Novartis to concentrate its global research efforts in Cambridge spurred other top pharmaceutical companies to do the same.

Local leadership is also well informed by faculty members at the area's universities who provide expertise, helping address community problems with data and thoughtful analysis. "There's a transmission line between what goes on in universities and how many academics inform and try and help improve the functioning of the city and regional government," Bluestone says. During the COVID-19 pandemic, for example, researchers at the Harvard T. H. Chan School of Public Health advised the city on the safety of school reopening and other issues.

There's also a civic culture, baked into the character of the region, that encourages participation. In August 1629, the members of the Massachusetts Bay Company signed an agreement in Cambridge, England, allowing the colonists moving to Massachusetts the right of local control over their affairs. John Winthrop and the Puritans carried this charter across the Atlantic, establishing self-governance when they arrived in 1630. They also created a "congregational" form of religion, in which each local congregation ran its own affairs rather than defer to the wishes of a monarch or pope. These were the first experiments in self-rule in English North America, distinctly different from the more hierarchical decision making that governed the Virginia Colony.

Today, Cambridge remains a politically aware and politically active community. In the November 2020 US presidential election, 75 percent of the city's 73,000 registered voters exercised their right to vote, well above the national average of 67 percent.[24] Some neighborhoods topped 80 percent voter turnout.

AND ONE WARNING: CITIES LIKE CAMBRIDGE CAN BECOME VICTIMS OF THEIR OWN SUCCESS UNLESS THEY ADDRESS CRITICAL ISSUES, SUCH AS INCOME INEQUALITY AND HOUSING AFFORDABILITY.   A growing chorus of people complain about the ruination of San Francisco.[25] The high-paying jobs of the tech industry drove up real estate prices and pushed out economic, racial, and ethnic diversity, making the city a far less interesting—and far more challenging—place to live. No American city lost more people in the second quarter of 2017 than San Francisco, real estate data shows.[26]

Cambridge, too, risks losing what has made it so attractive. Housing prices have skyrocketed. The average rental cost for a one-bedroom apartment in Cambridge rose 36 percent between 2010 and 2016, not including inflation.[27] One expansive, single-family house that sold for $280,000 in 1982 was worth $3.5 million by 2019—and was on the market for only four hours before it was snapped up by a foreign investor. Middle-class families who have lived in Cambridge for generations may make good money when they sell their family home, but then they can't afford to stay; teachers, firefighters, police officers, and other government employees can't live where they work.

Racism remains a problem, despite the city's reputation for tolerance.[28] There is tremendous income inequality between low-wage service jobs and high-wage positions in biotech and financial services, Bluestone notes. This split affects the school system— creating racial and class tensions—as well as the city's quality of life.

The biotech industry has historically skewed male and white. Only 10 percent of the companies started by MIT professors between 2000 and 2018 were founded by female professors, and only 14 percent of board seats at biotech firms were held by women.[29] In early 2020, professor emerita Nancy Hopkins, former MIT president Susan Hockfield, and bioengineering professor Sangeeta Bhatia formed the Boston

Biotech Working Group to try to foster more business participation by women through mentoring, offering externship opportunities for female professors, and circulating a pledge among local venture capital firms to address gender imbalance. The trade group Biotechnology Innovation Organization (BIO) made commitments in 2019 and 2020 to increase diversity among biotech boards and corporate leadership.[30] As we write this, it's too soon to know whether those efforts are making a difference.

The success of cities like Cambridge may also come at a cost to the larger American society, Richard Florida suggests in *The New Urban Crisis*. He warns that the clustering of talent and economic strengths in places like San Francisco, Boston, and other successful creative economies "generates a lopsided, unequal urbanism in which a relative handful of superstar cities, and a few elite neighborhoods within them, benefit while many other places stagnate or fall behind."[31] The answer, according to Florida and others, is not to destroy the cities that are the primary creators of wealth in our society—what he calls "the most powerful economic engines the world has ever seen"—but to distribute the rewards of this activity more equitably.[32] He suggests creating more affordable housing, addressing concentrations of poverty, reforming zoning codes that exclude people from participating in the urban economy, and turning low-wage service jobs into "family-supporting work."

Cambridge has at least tried to take some of these measures. The city has a well-funded public school system, spending more per pupil than all but a few communities in Massachusetts (already a state that ranks high in public education spending).[33] It rebuilt or renovated four schools and built two new public libraries between 2000 and 2019. Over roughly the same period, Cambridge allocated more than $154 million for affordable housing.[34] Addressing the issue of teen unemployment and workplace readiness, the Mayor's Summer Youth Employment Program has, since the 1970s, provided job opportunities to Cambridge high school students, most often a teen's first work experience. In recent years, the program has received more than a thousand applications each summer.[35]

For cities like Cambridge to continue to thrive, Bluestone says they will need a new kind of social innovation. Rather than inventing a new technology—speeding up the internet or improving gene editing—he says the next generation of great innovators must come up with solutions for housing, transportation, and health issues. "In addition to the great Thomas Edisons and Henry Fords, we need fine social innovators who are coming up with great ideas," Bluestone says. "They're going to be every bit as important to the future of this country."[36]

Maybe some of those great ideas will come from Cambridge.

NOTES

CHAPTER 1

1. "Auction Results: The Bay Psalm Book Sale," Sotheby's, November 26, 2013, http://www.sothebys.com/en/auctions/2013/the-bay-psalm-book-sale-n09039.html.

2. Finding Aid, H. A. Rey Papers, Ax 828, Special Collections & University Archives, University of Oregon Libraries, Eugene, Oregon, accessed March 21, 2020, http://archiveswest.orbiscascade.org/ark:/80444/xv65672.

3. Lois Lowry, "Lois Lowry: My House in Cambridge," personal website, accessed March 21, 2020, http://www.loislowry.com/index.php?option=com_content&view=article&id=75&Itemid=196.

4. Steven Pinker, "Steven Pinker, Department of Psychology, Harvard University," personal website, accessed March 21, 2020, https://stevenpinker.com/biocv.

5. Samuel Eliot Morison, *The Founding of Harvard College* (Cambridge, MA: Harvard University Press, 1968), 344–345.

6. Hope Mayo, in-person interview with the authors, Cambridge, MA, January 4, 2019.

7. Mayo, in-person interview with the authors, January 4, 2019.

8. Sidney A. Kimber, *The Story of an Old Press: An Account of the Hand Press Known as the Stephen Daye Press, upon Which Was Begun in 1638 the First Printing in British North America* (Cambridge, MA: University Press [ca. 1937]), 18–19.

9. American Library Association, "The Nation's Largest Libraries: A Listing by Volumes Held," ALA Library Fact Sheet 22, accessed July 19, 2019, http://www.ala.org/tools/libfactsheets/alalibraryfactsheet22.

10. "Little Free Library World Map," Little Free Library, accessed April 9, 2020, https://littlefreelibrary.org/ourmap.

11. Anne Bradstreet, Jeannine Hensley, and Adrienne Cecile Rich, *The Works of Anne Bradstreet* (Cambridge, MA: Belknap Press of Harvard University Press, 2010), xlvi.

12. Charlotte Gordon, in-person interview with the authors, Gloucester, MA, February 21, 2018.

13. Gordon, in-person interview with the authors, February 21, 2018.

14. Bradstreet, Hensley, and Rich, *The Works of Anne Bradstreet*, xlvii.

15. Charlotte Gordon, *Mistress Bradstreet: The Untold Life of America's First Poet* (New York: Little, Brown, 2005), 127.

16. Gordon, in-person interview with the authors, February 21, 2018.

17. Gordon, in-person interview with the authors, February 21, 2018.

18. For instance, Anne Hutchinson was banished from the Massachusetts Bay Colony in 1637 for daring to tell others how to practice their faith. Her trial was held in Cambridge.

19. "William Hill Brown," *Encyclopedia Britannica*, https://www.britannica.com/biography/William-Hill-Brown.

20. Bradstreet, Hensley, and Rich, *The Works of Anne Bradstreet*, xiv.

21. Gordon, *Mistress Bradstreet*, 19–20, 36–37.

22. Gordon, *Mistress Bradstreet*, 22–23.

23. Gordon, *Mistress Bradstreet*, 26. At St. Botolph's Church, listening to Cotton, Bradstreet would also have seen Hutchinson, then in her early thirties. Hutchinson, who was smart and outspoken, presented a more assertive, worldlier model of womanhood than Bradstreet's own mother.

24. Francis J. Bremer, "Cotton, John (1585–1652)," *Oxford Dictionary of National Geography*, accessed April 9, 2020, http://www.oxforddnb.com/view/10.1093/ref:odnb/9780198614128.001.0001/odnb-9780198614128-e-6416. The English of New England lived in a small world. John Cotton's son Seaborn later married Anne Bradstreet's daughter Dorothy; a daughter Maria married Increase Mather, the influential New England minister who became president of Harvard; Maria's son Cotton Mather also became a minister, a scientist, and a vocal opponent of the Salem witch trials.

25. Gordon, *Mistress Bradstreet*, 69–72.

26. Gordon, *Mistress Bradstreet*, 68–72.

27. Gordon, *Mistress Bradstreet*, 112.

28. Bradstreet, Hensley, and Rich, *The Works of Anne Bradstreet*, xlvi.

29. Bradstreet, Hensley, and Rich, *The Works of Anne Bradstreet*, xvii–xx.

30. Gordon, in-person interview with the authors, February 21, 2018.

31. Bradstreet, Hensley, and Rich, *The Works of Anne Bradstreet*, xxxii.

32. Bradstreet, Hensley, and Rich, *The Works of Anne Bradstreet*, xxxv.

33. Gordon, *Mistress Bradstreet*, 284.

34. Bradstreet, Hensley, and Rich, *The Works of Anne Bradstreet*, 240–241.

35. Matthew Gartner, "The Cultural Career of Longfellow's 'Paul Revere's Ride,'" in Christoph Irmscher and Robert Arbour, eds., *Reconsidering Longfellow* (Madison, NJ: Fairleigh Dickinson University Press, 2014), 124.

36. Edward Wagenknecht, *Henry Wadsworth Longfellow: Portrait of an American Humanist* (New York: Oxford University Press, 1966), 147.

37. Wagenknecht, *Henry Wadsworth Longfellow*, 6–7.

38. Irmscher and Arbour, *Reconsidering Longfellow*, 145.

39. Wagenknecht, *Henry Wadsworth Longfellow*, 8.

40. Irmscher and Arbour, *Reconsidering Longfellow*, 139.

41. Irmscher and Arbour, *Reconsidering Longfellow*, 150.

42. Charles C. Calhoun, *Longfellow: A Rediscovered Life* (Boston: Beacon Press, 2005), 198–199.

43. Christoph Irmscher, telephone interview with the authors, July 19, 2018.

44. "Longfellow's Elder Years," Maine Historical Society, accessed March 26, 2020, https://www.hwlongfellow.org/life_elder.shtml.

45. Irmscher, telephone interview with the authors, July 19, 2018.

46. Anna Christie, in-person interview with the authors, Longfellow House–Washington's Headquarters National Historic Site, July 5, 2018.

47. Irmscher, telephone interview with the authors, July 19, 2018.

48. Hillard's husband, George Stillman Hillard, who was a United States Commissioner, was responsible for issuing warrants for the capture of escaped enslaved people, per Kathryn Grover, "Site 25: George and Susan Hillard House," *Historic Resource Study Boston African American National Historic Site*, December 31, 2002, http://ininet.org/historic-resource-study-boston-african-american-national-histo.html?page=11.

49. Irmscher, telephone interview with the authors, July 19, 2018.

50. Irmscher, telephone interview with the authors, July 19, 2018.

51. Christie, in-person interview with the authors, July 5, 2018.

52. Irmscher and Arbour, *Reconsidering Longfellow*, 132.

53. Irmscher, telephone interview with the authors, July 19, 2018.

54. John Bartlett and Geoffrey O'Brien, *Bartlett's Familiar Quotations*, 18th ed. (Boston: Little, Brown, 2012), vii.

55. "John Bartlett, Author, Is Dead," *Cambridge Chronicle*, December 9, 1905, 15. Available at https://cambridge.dlconsulting.com/?a=d&d=Chronicle19051209-01.2.39&e=-------en-20--1--txt-txIN-------.

56. "Reminiscences of John Bartlett: Address of Joseph Willard," *Proceedings of the Cambridge Historical Society* 1 (April 24, 1906): 72, https://cambridgehistory.org/wp-content/uploads/2017/03/Proceedings-Volume-1-1905%E2%80%931906.pdf.

57. "*Bartlett's Familiar Quotations* Has Gone Digital," press release, Hachette Book Group, accessed March 1, 2020, http://www.bartlettsquotes.com.

58. Susan Wilson for the Boston History Collaborative, *The Literary Trail of Greater Boston*, rev. ed. (Beverly, MA: Commonwealth Editions, 2005), 103.

59. John Bartlett, "List: of books read: manuscript" [ca. 1900], MS Am 524, Houghton Library, Harvard University, Cambridge, MA, 9, accessed March 1, 2020, https://iiif.lib.harvard.edu /manifests/view/drs:42728812$1i.

60. Bartlett, "List: of books read," 8.

61. "Reminiscences of John Bartlett," 68.

62. Bartlett, "List: of books read," 10.

63. M. H. Morgan, "John Bartlett," in *Proceedings of the American Academy of Arts and Sciences*, vol. 41 (Boston: American Academy of Arts and Sciences, 1906), 842.

64. Bartlett, "List: of books read," 11.

65. "Reminiscences of John Bartlett," 70–71.

66. Wilson, *The Literary Trail of Greater Boston*, 102.

67. "Reminiscences of John Bartlett," 70.

68. Beginning in 1872, John Bartlett lived in a newly built house at 165 Brattle Street, about a quarter mile from Elmwood, the home of James Russell Lowell, per Christopher Hail, "Harvard/Radcliffe Online Historical Reference Shelf, Cambridge Buildings and Architects," accessed March 2, 2020, https://wayback.archive-it.org/5488/20170330145537/http://hul .harvard.edu/lib/archives/refshelf/cba/b.html.

69. "Reminiscences of John Bartlett," 84.

70. "Reminiscences of John Bartlett," 82.

71. Morgan, "John Bartlett," 843.

72. John Bartlett, ed., *Bartlett's Familiar Quotations*, 5th ed. (Boston: Little, Brown, 1868), vii, accessed February 12, 2018, https://archive.org/stream/familiarquotati03bartgoog#page/n10 /mode/2up.

73. "Reminiscences of John Bartlett," 73.

74. "Reminiscences of John Bartlett," 74.

75. "Reminiscences of John Bartlett," 86.

76. Morgan, "John Bartlett," 841.

77. John Gruesser, telephone interview with the authors, October 6, 2019.

78. Alisha Knight, telephone interview with the authors, October 3, 2019.

79. Ira Dworkin, "Biography of Pauline E. Hopkins (1859–1930)," The Pauline Elizabeth Hopkins Society, accessed July 6, 2019, http://www.paulinehopkinssociety.org/biography.

80. Brian Sweeney and Eurie Dahn, Project Directors, "The Digital Colored American Magazine," accessed April 9, 2020, https://coloredamerican.org/?page_id=373.

81. Pauline E. Hopkins, "A Primer of Facts Pertaining to the Early Greatness of the African Race and the Possibility of Restoration by Its Descendants—with Epilogue," in *Daughter of the*

Revolution: The Major Nonfiction Works of Pauline Hopkins, ed. Ira Dworkin (Piscataway, NJ: Rutgers University Press, 2007), 344.

82. Eurie Dahn, "Commentary," *Colored American Magazine* 6, no. 5 (March 1903), accessed July 6, 2019, https://coloredamerican.org/?page_id=373.

83. Kathleen Weiler, *Maria Baldwin's Worlds* (Amherst, MA: University of Massachusetts Press, 2019), 76.

84. Pauline E. Hopkins, "Letter to William Monroe Trotter, April 16, 1905," in *Daughter of the Revolution*, 238–248.

85. Hopkins, "Letter to William Monroe Trotter," 238–248.

86. Lois Brown, *Pauline Elizabeth Hopkins: Black Daughter of the Revolution* (Chapel Hill, NC: University of North Carolina Press. 2008), 166.

87. Brown, *Pauline Elizabeth Hopkins*, 196.

88. Brown, *Pauline Elizabeth Hopkins*, 178.

89. Hanna Wallinger, "On the Platform with Prominent Speakers," in *Pauline E. Hopkins: A Literary Biography* (Athens, GA: University of Georgia Press, 2005), 265–269, http://www.jstor.org/stable/j.ctt46nkjs.20.

90. Booker T. DeVaughn, "The Boston Literary and Historical Association: An Early 20th Century Example of Adult Education as Conducted by a Black Voluntary Association," *Lifelong Learning* 9, no. 4 (January 1, 1986): 11.

91. Brown, *Pauline Elizabeth Hopkins*, 3.

92. Brown, *Pauline Elizabeth Hopkins*, 525–531.

93. Ian Hamilton, *Robert Lowell: A Biography* (New York: Vintage Books, 1983), 303–304.

94. Lloyd Schwartz, telephone interview with the authors, September 10, 2019.

95. Hamilton, *Robert Lowell*, 391.

96. Hamilton, *Robert Lowell*, 308–309.

97. Lowell taught at Boston University in the 1950s and after moving to New York City in 1960, taught at Harvard from 1963 to 1970, commuting weekly to Cambridge from New York.

98. Lloyd Schwartz, telephone interview with the authors, September 10, 2019. Schwartz notes that in the time that he attended Lowell's seminars, Lowell had recently published *Near the Ocean* and was working on his *Notebook 1967–68*.

99. Kay R. Jamison, *Robert Lowell: Setting the River on Fire: A Study of Genius, Mania, and Character* (New York: Vintage, 2018), 3.

100. Jamison, *Robert Lowell*, 3–13.

101. Hamilton, *Robert Lowell*, 353.

102. Kathleen Spivack, *With Robert Lowell and His Circle* (Boston: Northeastern University Press, 2012), 72–76.

103. "About the Poetry Room," Woodberry Poetry Room, Harvard University, accessed March 18, 2020, https://library.harvard.edu/libraries/poetryroom#about.

104. "About Us," *Harvard Advocate*, accessed March 18, 2020, https://www.theharvardadvocate.com/about.

105. "The History of the Poets' Theatre," The Poets' Theatre, accessed March 19, 2020, https://www.poetstheatre.org/our-history.

106. Adam Kirsch, "The Young T. S. Eliot: A Rediscovery of an Emerging Poet," *Harvard Magazine*, July–August 2015, https://harvardmagazine.com/2015/07/the-young-t-s-eliot.

107. Malcolm Cowley, "cummings, e.e. (1894–1962)," Harvard Square Library, a digital library of Unitarian Universalist biographies, history, books, and media, abridged from "Cummings: One Man Alone," *Yale Review*, accessed March 19, 2020, https://www.harvardsquarelibrary.org/biographies/e-e-cummings-3.

108. Peter Davison, *The Fading Smile: From Robert Frost to Robert Lowell to Sylvia Plath* (New York: Knopf, 1994), 19.

109. Davison, *The Fading Smile*, 243–247.

110. Lenora P. Blouin, "May Sarton: A Poet's Life," accessed March 19, 2020, https://digital.library.upenn.edu/women/sarton/blouin-biography.html.

111. Megan Marshall, *Elizabeth Bishop: A Miracle for Breakfast* (New York: Houghton Mifflin Harcourt, 2017).

112. Elizabeth Bishop, "Five Flights Up," *New Yorker*, February 18, 1974, accessed March 19, 2020, https://www.newyorker.com/magazine/1974/02/25/five-flights-up.

113. Claudia Rankine, "Introduction," in Adrienne Rich, *Collected Poems, 1950–2012* (New York: Norton, 2016), xl.

114. "Seamus Heaney, 1939–2013," Poetry Foundation, accessed March 19 2020, https://www.poetryfoundation.org/poets/seamus-heaney.

115. Colleen Walsh, "Honoring, and Feeling, Heaney's Presence" *Harvard Gazette*, March 31, 2015.

116. Hilton Als, "Frank Bidart's Poetry of Saying the Unsaid," *New Yorker*, September 11, 2017.

117. James Sullivan, "At Home in a World of Words," *Boston Globe*, September 13, 2017.

118. "More about Louise Glück, Poet Laureate Consultant in Poetry, 2003–2004," Library of Congress, accessed March 21, 2020, https://www.loc.gov/poetry/more_gluck.html.

119. "Louise Glück, b. 1943," Poetry Foundation, accessed March 21, 2020, https://www.poetryfoundation.org/poets/louise-gluck.

120. Anders Olsson, "Bio-bibliography," Nobel Prize in Literature 2020, https://www.nobelprize.org/prizes/literature/2020/bio-bibliography.

CHAPTER 2

1. Andrea Estrada, "Dispatch from Fire Island: Manuscript Newly Transcribed by UCSB Thoreau Scholar Recounts Thoreau's Trip to the Site of a Shipwreck That Claimed the Life of Author and Journalist Margaret Fuller," *The Current*, University of California Santa Barbara, August 3, 2015, https://www.news.ucsb.edu/2015/015782/dispatch-fire-island.

2. Megan Marshall, in-person interview with the authors, Cambridge, MA, June 15, 2018.

3. Megan Marshall, *Margaret Fuller: A New American Life* (New York: First Mariner Books, 2013), 269.

4. Perry Miller, ed., *Margaret Fuller, American Romantic. A Selection from Her Writings and Correspondence* (Garden City, NY: Anchor Books, 1963), x.

5. Marshall, in-person interview with the authors, June 15, 2018.

6. Miller, *Margaret Fuller*, xiii.

7. Marshall, in-person interview with the authors, June 15, 2018.

8. Marshall, *Margaret Fuller*, 293.

9. Judith Thurman, "An Unfinished Woman," *New Yorker*, April 1, 2013.

10. Marshall, in-person interview with the authors, June 15, 2018.

11. Blanche Linden, *Silent City on a Hill: Landscapes of Memory and Boston's Mount Auburn Cemetery* (Columbus: Ohio State University Press, 1989), 208.

12. Linden, *Silent City on a Hill*, 309.

13. Bree Harvey and Meg Winslow, in-person interview with the authors, Mount Auburn Cemetery, Cambridge, MA, July 12, 2018.

14. Linden, *Silent City on a Hill*, 329.

15. Linden, *Silent City on a Hill*, 117.

16. Linden, *Silent City on a Hill*, 186.

17. Linden, *Silent City on a Hill*, 169–170.

18. Linden, *Silent City on a Hill*, 261.

19. Linden, *Silent City on a Hill*, 145.

20. Excerpt from Joseph Story, "Mount Auburn Consecrated," September 24, 1831, Friends of Mount Auburn, December 8, 2011, https://mountauburn.org/consecration.

21. Sylvia Wright Mitarachi, "The Life of Melusina Fay Peirce," final typescript, 51, Sylvia Wright Mitarachi Papers, 1834–1990, includes notes and work by Susana Robbins, 1990, MC 567, folder 1.1, Schlesinger Library, Radcliffe Institute, Harvard University, Cambridge, MA.

22. Mitarachi, "The Life of Melusina Fay Peirce," 63.

23. Mitarachi, "The Life of Melusina Fay Peirce," 6.

24. Mitarachi, "The Life of Melusina Fay Peirce," 1–100.

25. Dolores Hayden, *The Grand Domestic Revolution: A History of Feminist Designs for American Homes, Neighborhoods, and Cities* (Cambridge, MA: MIT Press, 1981), 1.

26. Dolores Hayden, telephone interview with the authors, August 14, 2019.

27. Mitarachi, "The Life of Melusina Fay Peirce," 72–73.

28. Mitarachi, "The Life of Melusina Fay Peirce," 53.

29. Mitarachi, "The Life of Melusina Fay Peirce," 102.

30. Mitarachi, "The Life of Melusina Fay Peirce," 140, 153.

31. Mitarachi, "The Life of Melusina Fay Peirce," 171.

32. Mitarachi, "The Life of Melusina Fay Peirce," 175–176.

33. Hayden, *The Grand Domestic Revolution*, 68.

34. Mitarachi, "The Life of Melusina Fay Peirce," 199.

35. Hayden, *The Grand Domestic Revolution*, 82.

36. Mitarachi, "The Life of Melusina Fay Peirce," 4, 221.

37. Hayden, *The Grand Domestic Revolution*, 80.

38. Mitarachi, "The Life of Melusina Fay Peirce," 213–217.

39. Mitarachi, "The Life of Melusina Fay Peirce," chap. 9, pp. 2–7.

40. Hayden, *The Grand Domestic Revolution*, 85.

41. "Maria Baldwin House: National Register of Historic Places Inventory—Nomination Form," National Park Service, accessed June 12, 2018, https://npgallery.nps.gov/pdfhost/docs/nhls/text/76000272.PDF.

42. Anthony W. Neal, "Maria Louise Baldwin: An Eminent Educator, Civic Leader, Speaker," *Boston Banner* (Boston, MA), May 2, 2013, 1, 19–20.

43. Dorothy B. Porter, "Maria Louise Baldwin, 1856–1922," *Journal of Negro Education* 21, no. 1 (Winter 1952): 94–96, http://www.jstor.org/stable/2965923.

44. N.D.B. Connolly, "To Remake the World: Slavery, Racial Capitalism, and Justice: This, Our Second Nadir," *Boston Review*, February 21, 2018, https://bostonreview.net/forum/remake-world-slavery-racial-capitalism-and-justice/n-d-b-connolly-our-second-nadir.

45. "Maria Baldwin House: National Register of Historic Places Inventory—Nomination Form."

46. W.E.B. Du Bois, ed., "Men of the Month: Maria Baldwin," *The Crisis*, National Association for the Advancement of Colored People, April 1917, 281.

47. Such as Hallie Q. Brown, *Homespun Heroines and Other Women of Distinction* (Xenia, OH: Aldine, 1926).

48. Lynn Boyd Porter, "Color Line Not Drawn and Merit Recognized," *Cambridge Chronicle*, August 29, 1903, 9.

49. Faustine Childress Jones-Wilson, Charles A. Asbury, D. Kamili Anderson, Sylvia M. Jacobs, and Margo Okazawa-Rey, *Encyclopedia of African-American Education* (Westport, CT: Greenwood Press, 1996), 41.

50. The school was renamed in her honor in 2002.

51. Porter, "Maria Louise Baldwin, 1856–1922," 94–96.

52. Daphne Abeel, "Maria Baldwin, 1856–1922: 'An Honor and a Glory,'" Cambridge Historical Society, accessed February 28, 2020, https://cambridgehistory.org/research/maria-baldwin-1856-1922-an-honor-and-a-glory.

53. Kathleen Weiler, telephone interview with the authors, January 2, 2020.

54. "Maria L. Baldwin Biography," Cambridge Public Schools, accessed June 12, 2018, https://baldwin.cpsd.us/about_our_school/baldwin_school_history/maria_l_baldwin_biography.

55. Susan D. Carle, *Defining the Struggle: National Organizing for Racial Justice, 1880–1915* (New York: Oxford University Press, 2013), 183.

56. Maude Thomas Jenkins, *The History of the Black Woman's Club Movement in America*, EdD diss., Teachers College, Columbia University (Ann Arbor: ProQuest Dissertations Publishing, 1984), 142.

57. "Dorothy B. Porter," in *Dictionary of American Negro Biography*, ed. Rayford Whittingham Logan and Michael R. Winston (New York: Norton, 1982), 21–22.

58. Charles Weldon Wadelington and Richard F. Knapp, *Charlotte Hawkins Brown and Palmer Memorial Institute: What One Young African American Woman Could Do* (Chapel Hill, NC: University of North Carolina Press, 1999), 29–30.

59. *The Radcliffe News*, January 14, 1921, 7.

60. Jenkins, *The History of the Black Woman's Club Movement in America*, 143–144.

61. Maria L. Baldwin, "The Changing Ideal of Progress," *Southern Workman* (Hampton, VA) 29, no. 1 (January 1900): 15–16.

62. Lynn Boyd Porter, "Color Line Not Drawn and Merit Recognized," 9.

63. Kathleen Weiler, *Maria Baldwin's Worlds: A Story of Black New England and the Fight for Racial Justice* (Amherst: University of Massachusetts Press, 2019), 175.

64. Pauline E. Hopkins, "Famous Women of the Negro Race VII: Educators," *Colored American Magazine* 5, no. 2 (June 1902): 127.

65. Zine Magubane, telephone interview with the authors, August 16, 2019.

66. Thomas C. Holt, "Du Bois, W.E.B.," in *African American National Biography*, ed. Henry Louis Gates Jr. and Evelyn Brooks Higginbotham (New York: Oxford University Press, 2008).

67. W. E. B. Du Bois, *The Souls of Black Folk* (New York: Pocket Books, 2005), viii.

68. Du Bois, *The Souls of Black Folk*, 3–8.

69. W. E. B. Du Bois, *The Autobiography of W. E. B. Du Bois* (New York: International Publishers, 1968), 133–134.

70. W. E. B. Du Bois, Henry Louis Gates, and Terri Hume Oliver, *The Souls of Black Folk: Authoritative Text, Contexts, Criticism* (New York: Norton, 1999), 191–192.

71. Du Bois, *Autobiography of W. E. B. Du Bois*, 132–153.

72. Du Bois, *Autobiography of W. E. B. Du Bois*, 135–136.

73. Du Bois, *Autobiography of W. E. B. Du Bois*, 146–150.

74. Penny Schwartz, "A New Kind of Massachusetts Wedding: In Cambridge, More Than 250 Same-Sex Couples Wait in Line to Be among the First to Apply for Marriage Licenses," *Jewish Advocate*, May 21, 2004.

75. Pam Belluck, "Massachusetts Arrives at Moment for Same Sex Marriage," *New York Times*, May 17, 2004, A16.

76. Nima Eshghi and Kate Eshghi, in-person interview with the authors, Lincoln, MA, September 8, 2019.

77. Steve Annear, "Cambridge City Hall Celebrates Same-Sex Marriage Ruling," *Boston Globe*, June 26, 2015.

78. GLBTQ Legal Advocates and Defenders, accessed February 17, 2020, http://www.glad.org/about/history.

79. Hillary Goodridge v. Department of Public Health, 440 Mass. 309, March 4, 2003–November 18, 2003, http://masscases.com/cases/sjc/440/440mass309.html.

80. Kimberly D. Richman, *License to Wed: What Legal Marriage Means to Same-Sex Couples* (New York: NYU Press, 2013), xx–xxi.

81. Carey Goldberg, "Vermont Gives Final Approval to Same-Sex Unions," *New York Times*, April 26, 2000.

82. Annear, "Cambridge City Hall Celebrates Same-Sex Marriage Ruling."

83. Victoria Whitley-Berry, "The 1st Legally Married Same-Sex Couple 'Wanted to Lead by Example,'" National Public Radio, May 17, 2019, www.npr.org/2019/05/17/723649385/the-1st-legally-married-same-sex-couple-wanted-to-lead-by-example.

84. Michael P. Norton, "On 14th Anniversary, Same-Sex Marriage Seen as 'Same-Old, Same-Old,'" *Cambridge Chronicle*, May 17, 2018, https://cambridge.wickedlocal.com/news/20180517/on-14th-anniversary-same-sex-marriage-seen-as-same-old-same-old.

85. Andrew Gelman, "Same-Sex Divorce Rate Not as Low as It Seemed," *Washington Post*, December 15, 2014. https://www.washingtonpost.com/news/monkey-cage/wp/2014/12/15/same-sex-divorce-rate-not-as-low-as-it-seemed.

86. Steve LeBlanc, "Numbers Show How Gay Marriage Has Fared in Massachusetts," Associated Press, June 9, 2015, https://www.masslive.com/news/index.ssf/2015/06/numbers_show_how_gay_marriage.html.

CHAPTER 3

1. O. Laville Stone, *History of Massachusetts Industries: Their Inception, Growth, and Success.* (Boston: S. J. Clarke Pub. Co., 1930), 773.

2. Gavin Weightman, *The Frozen Water Trade: A True Story* (New York: Hyperion, 2003), 21.

3. Jill Sinclair, *Fresh Pond: The History of a Cambridge Landscape* (Cambridge, MA: MIT Press, 2009), 34.

4. Weightman, *The Frozen Water Trade*, 27–32.

5. Scott Kirsner, "'Ice King' Frederic Tudor Was One Cool Character," *Boston Globe*, August 17, 2018.

6. Reid Mitenbuler, "The Stubborn American Who Brought Ice to the World," *The Atlantic*, February 5, 2013.

7. Carl Seaburg and Stanley Paterson, *The Ice King: Frederic Tudor and His Circle* (Boston: Massachusetts Historical Society and Mystic, CT: Mystic Seaport, 2003), 122.

8. Seaburg and Paterson, *The Ice King*, 122–123.

9. Weightman, *The Frozen Water Trade*, 109.

10. Sinclair, *Fresh Pond*, 36.

11. "Nathaniel Wyeth (1802–1856)," Oregon History Project, March 17, 2018, https://oregonhistoryproject.org/articles/biographies/nathaniel-wyeth-biography/#.XCGqnM9Kg_V.

12. Weightman, *The Frozen Water Trade*, 144.

13. Weightman, *The Frozen Water Trade*, 109.

14. Sinclair, *Fresh Pond*, 41–42.

15. Weightman, *The Frozen Water Trade*, 155.

16. *Cambridge on the Cutting Edge: Innovators and Inventions* (Cambridge, MA: Cambridge Historical Society, undated), 15.

17. Weightman, *The Frozen Water Trade*, 194.

18. Weightman, *The Frozen Water Trade*, 169.

19. Weightman, *The Frozen Water Trade*, 244.

20. "Elias Howe, Jr., Inventor of the Sewing Machine, 1819–1919: A Centennial Address," *Proceedings of the Cambridge Historical Society 14* (1919).

21. "Elias Howe, Jr., Inventor of the Sewing Machine."

22. James Parton, *History of the Sewing Machine* (Raleigh, NC: Howe Machine Company, 1867), 1. This forty-six-page document was published, possibly with Howe's involvement, as an insert in the *Atlantic Magazine*, essentially as an extended advertisement for the Howe Machine Co. Other sources lack the detail of this telling by writer James Parton or seem to derive most of their information from it.

23. Parton, *History of the Sewing Machine*, 3.

24. Parton, *History of the Sewing Machine*, 4.

25. "Elias Howe, Jr., Inventor of the Sewing Machine."

26. Philip G. Hubert Jr., *Men of Achievement: Inventors* (New York: Charles Scribner's Sons, 1893), 102–103.

27. Parton, *History of the Sewing Machine*, 5.

28. "Elias Howe, Jr., Inventor of the Sewing Machine."

29. "Obituary: Elias Howe," *New York Times*, October 5, 1867.

30. "Obituary: Elias Howe." The Howes are likely to have lived at 181 Main Street in Cambridge, a residential building and store that was owned by Fisher in 1844. The building was demolished in 1918.

31. Parton, *History of the Sewing Machine*, 6–7.

32. *The Odd Fellow's Companion: Devoted to the Interests of the Independent Order of Odd Fellows* 3, no. 7 (Columbus, OH: M. C. Lilley & Co., 1868), 386.

33. Parton, *History of the Sewing Machine*, 9.

34. "1846 Howe Jr.'s Sewing Machine Patent Model," National Museum of American History, accessed February 17, 2020, https://americanhistory.si.edu/collections/search/object/nmah_630930.

35. Parton, *History of the Sewing Machine*, 9.

36. Parton, *History of the Sewing Machine*, 20–24.

37. Hubert, *Men of Achievement*, 109.

38. "Obituary: Elias Howe," *New York Times*, October 5, 1867.

39. Parton, *History of the Sewing Machine*, 15.

40. Parton, *History of the Sewing Machine*, 20–24.

41. Hubert, *Men of Achievement*, 109.

42. "Obituary: Elias Howe," *New York Times*, October 5, 1867.

43. *The Bell Telephone: The Deposition of Alexander Graham Bell in the Suit Brought by the United States to Annul the Bell Patents* (Boston: American Bell Telephone Company, 1908), 129.

44. *The Bell Telephone: The Deposition of Alexander Graham Bell*, 129.

45. Edwin S. Grosvenor and Morgan Wesson, *Alexander Graham Bell: The Life and Times of the Man Who Invented the Telephone* (New York: Abrams, 1997), 49.

46. Christopher Beauchamp, *Invented by Law: Alexander Graham Bell and the Patent That Changed America* (Cambridge, MA: Harvard University Press, 2015), 35.

47. Grosvenor and Wesson, *Alexander Graham Bell*, 51.

48. Grosvenor and Wesson, *Alexander Graham Bell*, 67–72.

49. Beauchamp, *Invented by Law*, 68.

50. Beauchamp, *Invented by Law*, 38.

51. *The Bell Telephone: The Deposition of Alexander Graham Bell*, 132.

52. Grosvenor and Wesson, *Alexander Graham Bell*, 64.

53. "Gardiner Greene Hubbard," *Science* 6, no. 157 (December 31, 1897): 974–977.

54. Scott Kirsner, "A Factory in Cambridge Makes 14 Million Junior Mints a Day: Why Is No One Allowed Inside?," *Boston Globe*, May 4, 2018.

55. Alistair Birrell, "Dear America, We Invented All Your Candy. Love, Boston," Boston .com, October 22, 2014, accessed September 15, 2018, https://www.boston.com/culture/food /2014/10/22/dear-america-we-invented-all-your-candy-love-boston.

56. "Industrial and Old Home Week Edition," *Cambridge Chronicle*, July 27, 1907, 11.

57. Madeline Bilis, "A Brief History of Cambridge's Confectioner's Row," *Boston Magazine*, October 2016.

58. "George Close Company Building," National Register of Historic Places Registration Form, National Park Service, January 28, 2019, Massachusetts Cultural Resource Information System, accessed February 16, 2020, http://mhc-macris.net/Details.aspx?MhcId=CAM.1409.

59. *America's Unknown City: Cambridge, Massachusetts, 1630–1936* (Cambridge, MA: Harvard Trust Company, 1936), 9–10.

60. "Industrial and Old Home Week Edition," *Cambridge Chronicle*, July 27, 1907, 21.

61. Samira Kawash, *Candy: A Century of Panic and Pleasure* (New York: Faber & Faber, 2013), 154–164.

62. Kawash, *Candy*, 117–123.

63. "George Close Company Building," National Register of Historic Places Registration Form.

64. Kawash, *Candy*, 165.

65. Gus Rancatore, "Hello," email to the authors, March 23, 2020.

66. Gus Rancatore, telephone interview with the authors, March 24, 2020.

67. "George Close Company Building," National Register of Historic Places Registration Form.

68. "Industrial and Old Home Week Edition," 21.

69. Orra L. Stone, *History of Massachusetts Industries: Their Inception, Growth, and Success* (Boston, MA: S. J. Clarke Publishing, 1930), 820. See also "George Close Company Building," National Register of Historic Places Registration Form.

70. "Squirrel Brand: History," John B. Sanfilippo & Son, Inc., accessed March 26, 2020, http:// www.squirrelbrand.com/history.

71. "Deaths: Hollis G. Gerrish Candy Manufacturer," *Washington Post*, November 16, 1997.

72. "The History of Candy Making in Cambridge: Fox Cross Company," Cambridge Historical Society, last modified 2011, accessed September 15, 2018, https://cambridgehistory.org /candy/foxcross.html.

73. *America's Unknown City: Cambridge, Massachusetts, 1630–1936*, 9–10.

74. Katie MacDonald, "Automated Candy Production: 250 Massachusetts Avenue," Cambridge Historical Society, accessed September 15, 2018, https://cambridgehistory.org /innovation/Automatic%20Candy.

75. Eli Rosenberg, "Necco Wafer Factory Abruptly Shuts Down after Company Is Sold to Unknown Buyer," *Washington Post*, July 25, 2018.

76. "The History of Candy Making in Cambridge: Daggett Chocolate," Cambridge Historical Society, accessed September 15, 2018, https://cambridgehistory.org/candy/daggett.html.

77. "Fig Newtons," Casey's Videos, YouTube, posted July 6 2008, https://www.youtube .com/watch?v=UyI3IL46yq4.

78. Andrew Adam Newman, "Reminders That a Cookie Goes beyond the Fig," *New York Times*, May 1, 2012, B2.

79. National Register of Historic Places Registration Form, F. A. Kennedy Steam Bakery, June 10, 1988, accessed February 16, 2020, http://mhc-macris.net/Details.aspx?MhcId=CAM.1329.

80. Orra L. Stone, *History of Massachusetts Industries, Their Inception, Growth and Success* (Boston: S. J. Clarke Publishing Co., 1930), 84.

81. "Fig Newtons and the Kennedy Biscuit Company," Cambridge Historical Commission Archives and Library Blog, January 16, 2019, accessed February 16, 2020.

82. William Cahn, *Out of the Cracker Barrel: The Nabisco Story from Animal Crackers to Zuzus* (New York: Simon and Schuster, 1969), 101–103.

83. Christopher B. Daly, "Boston Suburb Makes a Date to Celebrate the Fig at Cookie Centennial," *Washington Post*, May 4, 1991.

84. "Fig Newtons and the Kennedy Biscuit Company."

85. David Saks, "The Company That's Keeping the Polaroid Legacy Alive," Bloomberg Business, April 11, 2016, https://www.bloomberg.com/features/2016-design/a/oskar-smolokowski.

86. Carl Johnson, "Edwin Land and Steve Jobs: Masters of Art and Science: What I Learned from the Polaroid's Founder Who Happened to Be the Apple CEO's Inspiration," *Ad Age*, October 14, 2011, https://adage.com/article/guest-columnists/polaroid-s-edwin-land-apple-s -steve-jobs/230446.

87. Robert Alter, telephone interview with the authors, March 9, 2020. Land's office was dismantled and destroyed. "It would have been a beautiful little museum," Alter says. "It's a shame because it's lost now."

88. Peter C. Wensberg, *Land's Polaroid, A Company and the Man Who Invented It* (Boston: Houghton, Mifflin, 1987), 20–34.

89. Victor McElheny, *Insisting on the Impossible: The Life of Edwin Land* (Reading, MA: Perseus Books, 1998), 47.

90. McElheny, *Insisting on the Impossible*, 68.

91. Wensberg, *Land's Polaroid*, 47–58.

92. Wensberg, *Land's Polaroid*, 71–74.

93. Wensberg, *Land's Polaroid*, 82.

94. McElheny, *Insisting on the Impossible*, 164.

95. Wensberg, *Land's Polaroid*, 99.

96. McElheny, *Insisting on the Impossible*, 198.

97. Milton Dentch, *Fall of an Icon, Polaroid after Edwin H. Land. An Insider's View of the Once Great Company* (Whitman, MA: Riverhaven Books, 2012), 44–45, 52.

98. Renee Garrelick, "Marian Stanley, Polaroid Corporation, Vice President of Emerging Markets, 140 Monument Street," interviewed May 16, 1996, Concord Oral History Program, https://concordlibrary.org/special-collections/oral-history/Stanley.

99. McElheny, *Insisting on the Impossible*, 199.

100. John Reuter, telephone interview with the authors, August 1, 2019.

101. McElheny, *Insisting on the Impossible*, 203.

102. Garrelick, "Marian Stanley."

103. Dan Corditz, "How Polaroid Bet Its Future on the SX-70," *Fortune*, January 1974, 83.

104. Robert Alter, telephone interview with the authors, March 9, 2020.

105. Peter W. Bernstein, "Polaroid Struggles to Get Back in Focus," *Fortune*, April 7, 1980, 68.

106. "A History of the Carter's Ink Company," Carter's Ink Company, Cambridge, MA, 1975, 1–5, Carter's Ink Collection, Cambridge Historical Commission.

107. E. Faulkner and Lucy Faulkner, "Let's Talk about Ink," *Bottles and Extras*, Spring 2003, 41–42, accessed February 17, 2020, https://www.fohbc.org/PDF_Files/Ink_Sp2003.pdf.

108. "Finding Aid, Carter's Ink Company Collection, September 2019," Cambridge Historical Commission, accessed February 17, 2020, https://www.cambridgema.gov/~/media/Files/historicalcommission/pdf/findingaids/fa_carterink_newoct2019.pdf?la=en.

109. "Industry in Cambridge: Carter's Ink," Cambridge Historical Society, accessed February 17, 2020, https://cambridgehistory.org/industry/cartersink.html.

110. Faulkner and Faulkner, "Let's Talk about Ink," 42.

111. "A History of the Carter's Ink Company," 10.

112. Francis Honn, "The Life and Times of Francis Honn," unpublished memoir obtained by the authors from members of the Honn family, 2018, 80–85.

113. Honn, "The Life and Times of Francis Honn," 80.

114. "Early Animated Suzy Q Pepsodent TV Commercial," VintageTVCommercials.com, YouTube, posted November 6, 2008, https://www.youtube.com/watch?v=iPsoxmXjtfc.

115. Honn, *The Life and Times of Francis Honn*, 85–86.

116. Hilary Greenbaum and Dana Rubinstein, "The Hand Held Highlighter," *New York Times Magazine*, January 20, 2012.

117. "Death Notice: Francis Jerome Honn," *New York Times*, July 24, 2016.

118. "A History of the Carter's Ink Company," 11.

119. "Highlighter," Wikipedia, https://en.wikipedia.org/wiki/Highlighter.

120. Jacob Hirschmann, "Hi-Lite of a Lifetime," *JCU (John Carroll University) Alumni Magazine*, March 7, 2016.

CHAPTER 4

1. "Science in Harvard's Teaching Cabinet, 1766–1820," Harvard Art Museums, accessed July 20, 2019, https://www.harvardartmuseums.org/visit/exhibitions/4916/the-philosophy -chamber-art-and-science-in-harvards-teaching-cabinet-1766-1820.

2. "Nobel Laureates," Harvard University, accessed March 26, 2020, https://www.harvard .edu/about-harvard/harvard-glance/honors/nobel-laureates.

3. "Gallery of Nobel laureates: MIT School of Humanities, Arts, and Social Sciences," Massachusetts Institute of Technology, accessed March 26, 2020, https://shass.mit.edu/about /honors/nobel-laureates.

4. "Honors and Awards Database," MIT Institutional Research, Office of the Provost, accessed March 27, 2020, http://ir.mit.edu/awards-honors.

5. I. Bernard Cohen, ed., *The Life and Scientific and Medical Career of Benjamin Waterhouse*, vol. 1 (New York: Arno Press, 1980), 18–21.

6. Cohen, *The Life and Scientific and Medical Career of Benjamin Waterhouse*, 1:25.

7. Philip Cash, *Dr. Benjamin Waterhouse: A Life in Medicine and Public Service (1754–1846)* (Sagamore Beach, MA: Science History Publications/USA, 2006), 10.

8. Cohen, *The Life and Scientific and Medical Career of Benjamin Waterhouse*, 1:175.

9. Cash, *Dr. Benjamin Waterhouse*, 24.

10. Cohen, *The Life and Scientific and Medical Career of Benjamin Waterhouse*, 1:176.

11. Cohen, *The Life and Scientific and Medical Career of Benjamin Waterhouse*, 1:56.

12. Cash, *Dr. Benjamin Waterhouse*, 59.

13. Cash, *Dr. Benjamin Waterhouse*, 86.

14. John B. Blake, "Benjamin Waterhouse, Harvard's First Professor of Physic," *Journal of Medical Education* 33, no. 11 (November 1958), reprinted in Cohen, *The Life and Scientific and Medical Career of Benjamin Waterhouse*, 2:775.

15. Cohen, *The Life and Scientific and Medical Career of Benjamin Waterhouse*, 1:144.

16. Cohen, *The Life and Scientific and Medical Career of Benjamin Waterhouse*, 1:93.

17. Cohen, *The Life and Scientific and Medical Career of Benjamin Waterhouse*, 1:106.

18. Robert H. Halsey, "How the President, Thomas Jefferson, and Doctor Benjamin Waterhouse Established Vaccination as a Public Health Procedure," paper presented before the section of Historical and Cultural Medicine, New York Academy of Medicine, 1936, as reprinted in Cohen, *The Life and Scientific and Medical Career of Benjamin Waterhouse*, 2:10.

19. Halsey, "How the President, Thomas Jefferson, and Doctor Benjamin Waterhouse Established Vaccination as a Public Health Procedure," 773.

20. John Blake, *Benjamin Waterhouse and the Introduction of Vaccination: A Reappraisal* (Philadelphia: University of Pennsylvania Press, 1957), 14.

21. Cohen, *The Life and Scientific and Medical Career of Benjamin Waterhouse*, 1:184.

22. "The Fight Over Inoculation during the 1721 Boston Smallpox Epidemic," Science in the News, Harvard Medical School, http://sitn.hms.harvard.edu/flash/special-edition-on -infectious-disease/2014/the-fight-over-inoculation-during-the-1721-boston-smallpox -epidemic.

23. Cohen, *The Life and Scientific and Medical Career of Benjamin Waterhouse*, 1:184.

24. Cash, *Dr. Benjamin Waterhouse*, 122.

25. Blake, *Benjamin Waterhouse and the Introduction of Vaccination*, 29.

26. Blake, *Benjamin Waterhouse and the Introduction of Vaccination*, 67.

27. Blake, *Benjamin Waterhouse and the Introduction of Vaccination*, 27–29.

28. Cohen, *The Life and Scientific and Medical Career of Benjamin Waterhouse*, 1:196. More details in Cohen, *The Life and Scientific and Medical Career of Benjamin Waterhouse*, 2:777–779.

29. Cohen, *The Life and Scientific and Medical Career of Benjamin Waterhouse*, 1:198.

30. "Smallpox," US Centers for Disease Control and Prevention, accessed April 7, 2020, https://www.cdc.gov/smallpox/index.html.

31. "Smallpox," World Health Organization, accessed April 7, 2020, http://www.who.int /csr/disease/smallpox/en.

32. Bessie Zaban Jones and Lyle Gifford Boyd, *The Harvard College Observatory: The First Four Directorships, 1839–1919* (Cambridge, MA: Belknap Press of Harvard University Press, 1971), 1–4.

33. Jonathan McDowell, in-person interview with the authors, Harvard-Smithsonian Center for Astrophysics, April 20, 2018.

34. Owen Jay Gingerich, in-person interview with the authors, Harvard-Smithsonian Center for Astrophysics, May 5, 2018.

35. "The Great Refractor," *Harvard Magazine*, May–June 2004, https://harvardmagazine .com/2004/05/the-great-refractor.html.

36. Edward S. Holden, *Memorials of William Cranch Bond, Director of the Harvard College Observatory, 1840–1859, and of His Son, George Phillips Bond, Director of the Harvard College Observatory, 1859–1965* (New York: Arno Press, 1980), 155.

37. Deborah Jean Warner, *Alvan Clark and Sons: Artists in Optics* (Washington, DC: Smithsonian Institution Press, 1968), 66.

38. McDowell, in-person interview with the authors, April 20, 2018.

39. Warner, *Alvan Clark and Sons*, 8.

40. Warner, *Alvan Clark and Sons*, 17.

41. Warner, *Alvan Clark and Sons*, 21.

42. Dava Sobel, *The Glass Universe: How the Ladies of the Harvard Observatory Took the Measure of the Stars* (New York: Viking, 2016), 89.

43. Sobel, *The Glass Universe*, 5–20.

44. "Harvard College Observatory," Harvard-Smithsonian Center for Astrophysics, accessed February 2, 2020, https://www.cfa.harvard.edu/hco/plates.html.

45. Sobel, *The Glass Universe*, 277–278.

46. Sobel, *The Glass Universe*, 152.

47. Jones and Boyd, *The Harvard College Observatory*, 444.

48. McDowell, in-person interview with the authors, April 20, 2018.

49. "Letter no. 2136," Darwin Correspondence Project, accessed January 25, 2020, https://www.darwinproject.ac.uk/letter/DCP-LETT-2136.xml.

50. Charles Darwin, *On the Origin of Species, or the Preservation of Favoured Races in the Struggle for Life* (London: John Murray, Albemarle Street, 1859), https://www.gutenberg.org/files/1228/1228-h/1228-h.htm.

51. A. Hunter Dupree, *Asa Gray: American Botanist, Friend of Darwin* (Baltimore, MD: Johns Hopkins University Press, 1988), 190–192.

52. Dupree, *Asa Gray*, 240–251.

53. James Dwight Dana and William Farlow, *Biographical Memoirs of Asa Gray* (Washington, DC: Government Printing Office, 1890), 755.

54. Dupree, *Asa Gray*, 134.

55. Dupree, *Asa Gray*, 116–117.

56. "Gray Herbarium," Harvard University Herbaria and Libraries, accessed July 14, 2018, https://huh.harvard.edu/pages/gray-herbarium-gh.

57. Dupree, *Asa Gray*, 121–122.

58. Dupree, *Asa Gray*, 212–213.

59. Dana and Farlow, *Biographical Memoirs of Asa Gray*, 769.

60. Dupree, *Asa Gray*, 203.

61. Dana and Farlow, *Biographical Memoirs of Asa Gray*, 753.

62. Louis Menand, *The Metaphysical Club: [A Story of Ideas in America]* (New York: Farrar, Straus and Giroux, 2007), 120.

63. Dupree, *Asa Gray*, 221.

64. Dupree, *Asa Gray*, 339.

65. Dana and Farlow, *Biographical Memoirs of Asa Gray*, 762.

66. Carolyn Ann Cobald, "The Rise of Alternative Bread Leavening Technologies in the Nineteenth Century," *Annals of Science* 75, no. 1 (2017): 21–39, doi: 10.1080/00033790.2017.1400100.

67. "Development of Baking Powder: National Historic Chemical Landmark," American Chemical Society, accessed June 17, 2018, https://www.acs.org/content/dam/acsorg/education/whatischemistry/landmarks/bakingpowder/rumford-baking-powder-commemorative-booklet.pdf.

68. Alvin Powell, "Bubble, Bubble—without Toil or Trouble," *Harvard Gazette*, April 2012, https://news.harvard.edu/gazette/story/2012/04/bubble-bubble-without-toil-or-trouble.

69. Charles L. Jackson, "Eben Norton Horsford," *Proceedings of the American Academy of Arts and Sciences* 28 (May 1892–May 1893): 343. https://www.jstor.org/stable/20020530?seq=3#page_scan_tab_contents.

70. "Development of Baking Powder," American Chemical Society.

71. Eben Norton Horsford, *The Theory and Art of Bread Making: A New Process without the Use of Ferment* (Cambridge, MA: Welch, Bigelow, and Company, 1861), 1–27.

72. "Development of Baking Powder," American Chemical Society.

73. Jackson, "Eben Norton Horsford," 342–343.

74. *In Memoriam. Eben Norton Horsford. July Twenty-Seventh, 1818. January First, 1893* (Wellesley, MA: Wellesley College, 1893), 14, https://archive.org/details/inmemoriamebenno00well/page/n8/mode/2up.

75. Eben Norton Horsford, *The Discovery of the Ancient City of Norumbega. Communication to the President and Council of the American Geographical Society at their Special Session in Watertown, November 21, 1889* (Boston: Houghton, Mifflin, 1889), 5.

76. Unsigned review of *The Defences of Norumbega* by Eben Norton Horsford, *Science* 17, no. 435 (June 5, 1891): 319.

77. Jackson, "Eben Norton Horsford," 343.

78. William James, "Louis Agassiz," *Science* 5, no. 112 (February 19, 1897): 286–288.

79. W. James, "Louis Agassiz: Words Spoken by Professor William James at the Reception of the American Society of Naturalists by the President and Fellows of Harvard College, at Cambridge, on December 30, 1896," Harvard University, Cambridge, MA, 1897, 9. Available at https://doi.org/10.5962/bhl.title.48644.

80. "Louis Agassiz," *Scientific American* 30, no. 1 (January 3, 1874): 2.

81. Christoph Irmscher, *Louis Agassiz: Creator of American Science* (Boston: Houghton Mifflin Harcourt, 2013), 65.

82. Edward Lurie, "Editor's Introduction," in Jean Louis Rodolphe Agassiz, *Essay on Classification*, ed. Edward Lurie (Cambridge, MA: Belknap Press of Harvard University Press, 1962), xiv.

83. "Obituary: Louis Agassiz," *American Naturalist* 32, no. 375 (March 1898): 157.

84. Louis Menand, *The Metaphysical Club* (New York: Farrar, Straus & Giroux, 2001), 100.

85. "Obituary: Louis Agassiz," 156.

86. Lurie, "Editor's Introduction," ix.

87. "Obituary: Louis Agassiz," 155.

88. Irmscher, *Louis Agassiz*, 108–110.

89. "Mission," National Academy of Sciences, accessed March 28, 2020, http://www .nasonline.org/about-nas/mission.

90. "Louis Agassiz," *Scientific American*, 2.

91. Irmscher, *Louis Agassiz*, 285.

92. Irmscher, *Louis Agassiz*, 2.

93. Menand, *The Metaphysical Club*, 111–113.

94. Irmscher, *Louis Agassiz*, 225.

95. Irmscher, *Louis Agassiz*, 268.

96. Brian Katz and Ewart Wetherill, "Fogg Art Museum Lecture Room: A Calibrated Recreation of the Birthplace of Room Acoustics," *Journal of the Acoustical Society of America* 120 (2005): 2191, doi: 10.1121/1.4787033.

97. Wallace C. Sabine, *Reprints from the American Architect on Architectural Acoustics. Part I—Reverberation* (Cambridge, MA, 1900), 9.

98. Leo L. Beranek, "The Notebooks of Wallace C. Sabine," *Journal of the Acoustical Society of America* 61, no. 629 (1977): 630–635, doi: 10.1121/1.381348.

99. Leo L. Beranek and John W. Kopec, "Wallace C. Sabine, Acoustical Consultant," *Journal of the Acoustical Society of America* 69, no. 1 (1981): 2, doi: 10.1121/1.385339.

100. Sabine, *Reprints from the American Architect*, 10.

101. Beranek, "The Notebooks of Wallace C. Sabine," 633.

102. D. Murray Campbell, "Lord Rayleigh: A Master of Theory and Experiment in Acoustics," *Acoustical Science and Technology*, 28, no. 4 (2007), https://www.jstage.jst.go.jp/article /ast/28/4/28_4_215/_pdf/-char/en.

103. Benjamin Markham, in-person interview with the authors, Cambridge, MA, July 11, 2018.

104. Carl Rosenberg, in-person interview with the authors, Cambridge, MA, July 11, 2018.

105. Beranek, "The Notebooks of Wallace C. Sabine," 634.

106. Beranek, "The Notebooks of Wallace C. Sabine," 636.

107. Sabine, *Reprints from the American Architect*, 24.

108. Sabine, *Reprints from the American Architect*, 13.

109. Beranek and Kopec, "Wallace C. Sabine, Acoustical Consultant," 1.

110. Beranek, "The Notebooks of Wallace C. Sabine," 633.

111. Beranek and Kopec, "Wallace C. Sabine, Acoustical Consultant," 8.

112. Beranek, "The Notebooks of Wallace C. Sabine," 632.

113. W. D. Orcutt, *Wallace Clement Sabine: A Study in Achievement* (Norwood, MA: Plimpton Press, 1933), viii.

114. Orcutt, *Wallace Clement Sabine*, 8–10.

115. Orcutt, *Wallace Clement Sabine*, 19–27.

116. Orcutt, *Wallace Clement Sabine*, 38.

117. Leo L. Beranek, "The Personal Papers of Wallace C. Sabine," *Journal of the Acoustical Society of America* 125 (2009): 3793, doi: 10.1121/1.3125326.

118. Beranek and Kopec, "Wallace C. Sabine, Acoustical Consultant," 16.

CHAPTER 5

1. Douglas Brinkley, "Eisenhower: His Farewell Speech as President Inaugurated the Spirit of the 1960s," *American Heritage Magazine* 52, no. 6 (September 2001), https://web.archive.org/web/20060323001947/http://www.americanheritage.com/articles/magazine/ah/2001/6/2001_6_58.shtml.

2. US Army Center of Military History, "Washington Takes Command of Continental Army in 1775," US Army, April 15, 2016, http://www.army.mil/article/40819/Washington_takes_command_of_Continental_Army_in_1775.

3. Michael Kenney and Kathleen Rawlins, "'To Protect the Union': Civil War History in Central Square," Cambridge Historical Commission, Cambridge, MA, January 2012, https://www.cambridgema.gov/~/media/files/historicalcommission/pdf/companyc.pdf?la=en.

4. "Technical Reports and Standards: The Office of Scientific Research and Development (OSRD) Collection," Science Reference Services, Library of Congress, November 18, 2015, https://www.loc.gov/rr/scitech/trs/trsosrd.html.

5. "The Salvador E. Luria Papers: Politics, Science and Social Responsibility," Profiles in Science, US National Library of Medicine, https://profiles.nlm.nih.gov/spotlight/ql/feature/politics.

6. Robert Conn, "Introductory Comments," A Symposium on the Occasion of the 75th Anniversary of Vannevar Bush's *Science: The Endless Frontier*, The National Academies of Sciences, Engineering, and Medicine, February 26, 2020, https://nationalacademies.org/next-75-years-in-science.

7. Robert Buderi, *The Invention That Changed the World: How a Small Group of Radar Pioneers Won the Second World War and Launched a Technical Revolution* (New York: Simon & Schuster, 1996), 471.

8. Gregg Pascal Zachary, *Endless Frontier: Vannevar Bush, Engineer of the American Century* (New York: Free Press, 1997), 34.

9. Vannevar Bush, *Pieces of the Action: The Personal Record of Sixty Event-Filled Years by the Distinguished Scientist Who Took an Active and Decisive Part in Shaping Them* (New York: Morrow, 1970), 74.

10. Raytheon became a major defense contractor during World War II, mass producing magnetrons, the key component in the microwave radar sets invented at MIT's Radiation Lab. Jerome Wiesner, "Vannevar Bush, March 11, 1890–June 28, 1974," in *Biographical Memoirs*, vol. 50 (Washington, DC: National Academies Press, 1979), https://doi.org/10.17226/573.

11. David Kaiser, ed., *Becoming MIT: Moments of Decision* (Cambridge, MA: MIT Press, 2012), 73–75.

12. "Technical Reports and Standards: The Office of Scientific Research and Development (OSRD) Collection," 4–5.

13. Zachary, *Endless Frontier*, 5.

14. Bush, *Pieces of the Action*, 1–8.

15. Buderi, *The Invention That Changed the World*, 104.

16. John E. Burchard, *Q.E.D.: MIT in World War II* (Cambridge, MA: Technology Press, 1946), 221.

17. Karl T. Compton, "President's Report 1944–1945," *Massachusetts Institute of Technology Bulletin* 81, no. 1 (1945): 12–17.

18. Burchard, *Q.E.D.: MIT in World War II*, 216.

19. Buderi, *The Invention That Changed the World*, 42.

20. Buderi, *The Invention That Changed the World*, 49.

21. Buderi, *The Invention That Changed the World*, 45.

22. Burchard, *Q.E.D.: MIT in World War II*, 223.

23. Buderi, *The Invention That Changed the World*, 98.

24. Burchard, *Q.E.D.: MIT in World War II*, 220–226.

25. "Radio Research Laboratory, 1942–1946," Collection of Historical Scientific Instruments, Harvard University, accessed March 7, 2020, http://waywiser.fas.harvard.edu/people/8725/radio-research-laboratory;jsessionid=693DDC9A31CA842ACD23E27528B59314.

26. Monroe Singer, "Harvard Radio Research Lab Developed Countermeasures against Enemy Defenses," *Harvard Crimson*, November 30, 1945, https://www.thecrimson.com/article/1945/11/30/harvard-radio-research-lab-developed-countermeasures.

27. "Radio Research Laboratory, 1942–1946."

28. "MIT Rad Lab," Google Arts & Culture, https://artsandculture.google.com/exhibit/mit-rad-lab-u-s-national-archives/_ALSxrIyTPWDJw?hl=en.

29. Burchard, *Q.E.D.: MIT in World War II*, 222.

30. Nobel Prize–winners in physics or chemistry who spent portions of their professional careers at the MIT Radiation Lab include the following (with year of Nobel award in parentheses): I. I. Rabi (1944), Edwin M. McMillan (1951), Edward Mills Purcell (1952), Julian Schwinger (1965), Hans Bethe (1967), Luis Walter Alvarez (1968), John Hasbrouck Van Vleck (1977), and Norman Foster Ramsey Jr. (1989).

31. Compton, "President's Report 1944–1945," 8.

32. Buderi, *The Invention That Changed the World*, 256.

33. Don Murray, "Percy Spencer and His Itch to Know," *Reader's Digest*, August 1958, 114.

34. Estelle Jussim, *Stopping Time: The Photographs of Harold Edgerton* (New York: Abrams, 1987), 35.

35. J. Kim Vandiver, in-person interview with the authors, Cambridge, MA, February 25, 2019.

36. Jussim, *Stopping Time*, 35.

37. Jussim, *Stopping Time*, 35.

38. A 1935 General Radio catalog indicates the availability of Model 548-A, the Edgerton stroboscope. "Catalog H, General Radio Company, Cambridge Mass. April 1935," accessed March 7, 2020, https://www.ietlabs.com/pdf/GR_Catalog/GenRad_CatH_1935.pdf.

39. Kate Flint, "Victorian Flash," *Journal of Victorian Culture* 23, no. 4 (October 2018): 481–489, https://academic.oup.com/jvc/article/23/4/481/5054041.

40. Jussim, *Stopping Time*, 36, 145.

41. Vandiver, in-person interview with the authors, February 25, 2019.

42. Vandiver, in-person interview with the authors, February 25, 2019.

43. "Flash: Photographs by Harold Edgerton from the Whitney's Collection, Mar 30–July 15, 2018," Whitney Museum of American Art, accessed March 7, 2020, https://whitney.org/Exhibitions/HaroldEdgerton.

44. Vandiver, in-person interview with the authors, February 25, 2019.

45. Vandiver, in-person interview with the authors, February 25, 2019.

46. "Strobe Project Laboratory," Edgerton Center, Massachusetts Institute of Technology, Cambridge, MA, https://edgerton.mit.edu/courses/strobe-project-laboratory.

47. Robert M. Neer, *Napalm, An American Biography* (Cambridge, MA: Belknap Press of Harvard University Press, 2013), 147.

48. Traci Tong, producer, "How the Vietnam War's Napalm Girl Found Hope after Tragedy," The World, PRX and WGBH radio, February 21, 2018, https://www.pri.org/stories/2018-02-21/how-vietnam-wars-napalm-girl-found-hope-after-tragedy.

49. Marshall Gates, "Louis Frederick Fieser," in *Biographical Memoirs*, vol. 65, chap. 7 (Washington, DC: National Academies Press, 1994), 160–167, https://www.nap.edu/read/4548/chapter/8.

50. Louis Fieser, *The Scientific Method: A Personal Account of Unusual Projects in War and in Peace* (New York: Reinhold, 1964), 9.

51. Fieser, *The Scientific Method*, 13–16.

52. Fieser, *The Scientific Method*, 21.

53. Fieser, *The Scientific Method*, 23.

54. Fieser, *The Scientific Method*, 23.

55. Neer, *Napalm*, 32.

56. Neer, *Napalm*, 32.

57. "Patent 2,606,107 Incendiary Gels, Louis F. Fieser, Belmont, Mass., Assignor to the United States of America as Represented by the Secretary of War Application November 1, 1943, Serial No. 508,63," United States Patent Office, Washington, DC.

58. "Patent 2,606,107 Incendiary Gels."

59. Fieser, *The Scientific Method*, 32.

60. Neer, *Napalm*, 43.

61. Neer, *Napalm*, 72.

62. William W. Ralph, "Improvised Destruction: Arnold, LeMay, and the Firebombing of Japan," *War in History* 13, no. 4 (2006): 495–522, http://www.jstor.org/stable/26061697.

63. Ralph, "Improvised Destruction," 496.

64. Neer, *Napalm*, 86.

65. Neer, *Napalm*, 132.

66. Nicholas Lemann, "Napalm's Daddy 31 Years Later," *Harvard Crimson*, October 12, 1973, https://www.thecrimson.com/article/1973/10/12/napalms-daddy-31-years-later-pin.

67. Jeff Abramson, "U.S. Incendiary-Weapons Policy Rebuffed," *Arms Control Today*, March 31, 2010, Arms Control Association, https://www.armscontrol.org/act/2010-03/us-incendiary-weapons-policy-rebuffed.

68. International Committee of the Red Cross, "Protocol on Prohibitions or Restrictions on the Use of Incendiary Weapons (Protocol III). Geneva, 10 October 1980," https://ihl

-databases.icrc.org/ihl/INTRO/515. On September 23, 2008, the United States Senate ratified the Protocol on Prohibitions or Restrictions on the Use of Incendiary Weapons, commonly called Protocol III, part of an international agreement known as the Convention on Certain Conventional Weapons (CCW).

69. Abramson, "U.S. Incendiary-Weapons Policy Rebuffed": "In prepared remarks at an April 15, 2008, Senate Foreign Relations Committee hearing that discussed the protocol, Charles Allen, deputy general counsel for international affairs at the Department of Defense, reiterated that 'incendiary weapons are the only weapons that can effectively destroy certain counter-proliferation targets such as biological weapons facilities, which require high heat to eliminate biotoxins.'"

70. "An Overdue Review: Addressing Incendiary Weapons in the Contemporary Context," Human Rights Watch, November 20, 2017, https://www.hrw.org/news/2017/11/20/overdue-review-addressing-incendiary-weapons-contemporary-context.

71. "An Overdue Review."

72. Charles Stark Draper, "The Evolution of Aerospace Guidance Technology at the Massachusetts Institute of Technology, 1935–1951: A Memoir," paper presented at the Fifth History Symposium of the International Academy of Astronautics, Brussels, Belgium, September 1971, in *Essays on the History of Rocketry and Astronautics: Proceedings of the Third through Sixth History Symposia of the International Academy of Astronautics*, vol. 2 (Washington, DC: NASA, 1971), 220, https://ntrs.nasa.gov/search.jsp?R=19770026118.

73. Draper, "The Evolution of Aerospace Guidance Technology at the Massachusetts Institute of Technology," 221.

74. Hugh Blair-Smith, telephone interview with the authors, August 30, 2019.

75. Kathleen Granchelli, ed., *Draper Laboratory: 40 Years as an Independent R&D Institution, 80 Years of Outstanding Innovations and Service to the Nation* (Cambridge, MA: Charles Stark Draper Laboratory, 2013), 9.

76. Granchelli, *Draper Laboratory*, 13.

77. Draper, "The Evolution of Aerospace Guidance Technology at the Massachusetts Institute of Technology."

78. Granchelli, *Draper Laboratory*, 15.

79. Blair-Smith, telephone interview with the authors, August 30, 2019.

80. David G. Hoag, *The History of Apollo On-Board Guidance, Navigation and Control* (Cambridge, MA: Charles Stark Draper Laboratory, 1976), 4.

81. Hoag, *The History of Apollo On-Board Guidance, Navigation and Control*, 5.

82. Don Eyles, "Tales from the Lunar Module Guidance Computer," paper presented to the Twenty-seventh Annual Guidance and Control Conference of the American Astronautical Society (AAS), Breckenridge, CO, February 6, 2004, https://www.doneyles.com/LM/Tales.html.

83. Margaret Hamilton, telephone interview with the authors, September 12, 2019.

84. Hamilton, telephone interview with the authors, September 12, 2019.

85. Lance Drane, telephone interview with the authors, August 30, 2019.

86. Julie Duke, "Slight Rebellion off Mass. Ave.," *Technology Review*, August 19, 2014, https://www.technologyreview.com/s/529716/slight-rebellion-off-mass-ave.

87. John Chute, John Lindsay, and Jay Mccleod, "Demonstration at Draper Lab: The Motivations Behind a Protest," *Harvard Crimson*, April 30, 1981, https://www.thecrimson.com/article/1981/4/30/demonstration-at-draper-lab-pifourteen-protestors.

88. Web of Science: Arts & Humanities Citation Index, Web of Science, Clarivate, accessed December 30 2020, https://clarivate.com/webofsciencegroup/solutions/webofscience-arts-and-humanities-citation-index.

89. Samuel Jay Keyser, email to the authors, April 19, 2020.

90. "Chomsky.Info: Recent Updates," accessed March 27, 2020, https://chomsky.info/updates.

91. "Linguistic Books by Noam Chomsky," Department of Linguistics, Massachusetts Institute of Technology, accessed March 2, 2019, http://linguistics.mit.edu/user/chomsky.

92. Samuel Jay Keyser, in-person interview with the authors, Cambridge, MA, January 15, 2020.

93. James Peck, ed., *The Chomsky Reader* (Pantheon Books, New York, 1987), 5.

94. Peck, *The Chomsky Reader*, 15.

95. Robert F. Barsky, *Noam Chomsky: A Life of Dissent* (Cambridge, MA: MIT Press, 1997), 87.

96. Gary Marcus, "Happy Birthday, Noam Chomsky," *New Yorker*, December 6, 2012, https://www.newyorker.com/news/news-desk/happy-birthday-noam-chomsky.

97. "Noam Chomsky—Is There a Form of Government We Can Trust?," National Press Club of Australia, YouTube, January 11, 2019, https://www.youtube.com/watch?v=7rrox1wdqzI.

98. Noam Chomsky, "Yugoslavia: Peace, War, and Dissolution," PM Press, Oakland, CA, April 10, 2018.

99. Noam Chomsky, *9–11: Was There an Alternative?* (New York: Seven Stories Press, 2001).

100. "Chomsky Echoes Prominent Israeli, Warns of the Rise of 'Judeo-Nazi Tendencies' in Israel," *Middle East Monitor*, November 12, 2018, https://www.middleeastmonitor.com/20181112-chomsky-echoes-prominent-israeli-warns-of-the-rise-of-judeo-nazi-tendencies-in-israel.

101. George Yancy, "Noam Chomsky: On Trump and the State of the Union," *New York Times*, July 5, 2017, https://www.nytimes.com/2017/07/05/opinion/noam-chomsky-on-trump-and-the-state-of-the-union.html.

CHAPTER 6

1. M. Mitchell Waldrop, "Claude Shannon: Reluctant Father of the Digital Age," *MIT Technology Review*, July 1, 2001, https://www.technologyreview.com/s/401112/claude-shannon-reluctant-father-of-the-digital-age.

2. Jimmy Soni and Rob Goodman, *A Mind at Play: How Claude Shannon Invented the Information Age* (New York: Simon and Schuster, 2017), 33.

3. Eugene Chiu, Jocelyn Lin, Brok Mcferron, Noshirwan Petigara, and Satwiksai Seshasai, "Mathematical Theory of Claude Shannon: A Study of the Style and Context of His Work up to the Genesis of Information Theory," class paper, 6.933J/STS.420J The Structure of Engineering Revolutions, Massachusetts Institute of Technology, Fall 2001, p. 31, http://web.mit.edu/6.933/www/Fall2001/Shannon1.pdf.

4. Claude Shannon, "A Symbolic Analysis of Relays and Switching Circuits," *AIEE Transactions* 57 (1938): 722–723.

5. Chiu et al., "Mathematical Theory of Claude Shannon," 22–33.

6. Soni and Goodman, *A Mind at Play*, 81.

7. "Claude Shannon," *Physics Today*, April 30, 2018, doi:10.1063/PT.6.6.20180430a.

8. Soni and Goodman, *A Mind at Play*, 165.

9. Robert Gallager, telephone interview with the authors, July 31, 2019.

10. Chiu et al., "Mathematical Theory of Claude Shannon," 60.

11. Soni and Goodman, *A Mind at Play*, 223–225.

12. Chiu et al., "Mathematical Theory of Claude Shannon," 60.

13. Glenn Rifkin, "Leo Beranek, Acoustics Designer and Internet Pioneer, Dies at 102," *New York Times*, October 17, 2016.

14. David Walden and Raymond Nickerson, eds., *A Culture of Innovation: Insider Accounts of Computing and Life at BBN* (East Sandwich, MA: Waterside Publishing, 2011), 23, https://walden-family.com/bbn/frontmatter.pdf.

15. Walden and Nickerson, *A Culture of Innovation*, 72.

16. Rifkin, "Leo Beranek, Acoustics Designer and Internet Pioneer, Dies at 102."

17. J. C. R. Licklider, "Man-Computer Symbiosis," *IRE Transactions on Human Factors in Electronics* HFE-1 (March 1960): 4–11, accessed June 25, 2018, http://groups.csail.mit.edu/medg/people/psz/Licklider.html.

18. Walden and Nickerson, *A Culture of Innovation*, 14–29.

19. Barry M. Leiner, Vinton G. Cerf, David D. Clark, Robert E. Kahn, Leonard Kleinrock, Daniel C. Lynch, Jon Postel, Lawrence G. Roberts, and Stephen Wolff, "A Brief History of the

Internet: Internet Hall of Fame," The Internet Society, accessed June 25, 2018, https://www
.internethalloffame.org/brief-history-internet.

20. Rifkin, "Leo Beranek, Acoustics Designer and Internet Pioneer, Dies at 102."

21. Robert E. Kahn, Oral history interview with Robert E. Kahn, Charles Babbage Institute,
1990, retrieved from the University of Minnesota Digital Conservancy, http://hdl.handle
.net/11299/107387.

22. Hiawatha Bray and Bryan Marquard, "Leo Beranek, 102, Pioneer Who Unlocked Mysteries of Acoustics," *Boston Globe*, October 13, 2016.

23. Dag Spicer, "Raymond Tomlinson: Email Pioneer, Part 1," *IEEE Annals of the History of
Computing* (April–June 2016) 75–79.

24. Rifkin, "Leo Beranek, Acoustics Designer and Internet Pioneer, Dies at 102."

25. Spicer, "Raymond Tomlinson: Email Pioneer, Part 1," 76.

26. Spicer, "Raymond Tomlinson: Email Pioneer, Part 1," 78.

27. S. L. Mathison, L. G. Roberts, and P. M. Walker, "The History of Telenet and the Com-
mercialization of Packet Switching in the U.S.," *IEEE Communications Magazine* 50, no. 5 (May
2012): 28–45, doi: 10.1109/MCOM.2012.6194380.

28. Spicer, "Raymond Tomlinson: Email Pioneer, Part 1," 72.

29. "Internet Hall of Fame: Raymond Tomlinson," The Internet Society, accessed June 25,
2018, https://internethalloffame.org/inductees/raymond-tomlinson.

30. William Grimes, "Raymond Tomlinson, Who Put the @ Sign in Email, Is Dead at 74,"
New York Times, March 7, 2016.

31. E. M. Rogers, "Claude Shannon's Cryptography Research during World War II and the
Mathematical Theory of Communication," *1994 Proceedings of IEEE International Carnahan
Conference on Security Technology*, Albuquerque, NM, USA (New York: IEEE, 1994), 1–5, doi:
10.1109/CCST.1994.363804.

32. Graham P. Collins, "Claude E. Shannon: Founder of Information Theory," *Scientific American*, October 14, 2002, https://www.scientificamerican.com/article/claude-e-shannon-founder.

33. "Whitfield Diffie," A. M. Turing Award, 2015, Association for Computing Machinery,
accessed June 26, 2018, https://amturing.acm.org/award_winners/diffie_8371646.cfm.

34. Adapted from a course description prepared by Leonid Grinberg for 6.045, Introduction to
Cryptography and RSA (as taught by Scott Aaronson), Massachusetts Institute of Technology,
Spring 2011, accessed June 27, 2018, https://ocw.mit.edu/courses/electrical-engineering-and
-computer-science/6-045j-automata-computability-and-complexity-spring-2011/lecture
-notes/MIT6_045JS11_rsa.pdf.

35. "Whitfield Diffie," A. M. Turing Award.

36. Leonard Adleman, interview with the authors, August 18, 2019.

37. "Len Adleman, 2002 ACM Turing Award Recipient," interviewed by Hugh Williams, August 18, 2016, Association for Computing Machinery, YouTube, October 25, 2016, https://www.youtube.com/watch?v=K06hOhABP-Y.

38. R. L. Rivest, A. Shamir, and L. Adleman, "A Method for Obtaining Digital Signatures and Public-Key Cryptosystems," *Communications of the ACM* 26, no. 1 (January 1983): 96–99.

39. "Oral History of Ronald L. Rivest," interviewed by Roy Levin, December 6, 2016, Computer History Museum, YouTube, July 3, 2017, https://www.youtube.com/watch?v=gQJmqQrcazU.

40. "The Birth of the Web," CERN (European Organization for Nuclear Research), accessed June 27, 2018, https://home.cern/topics/birth-web.

41. Dan Bricklin, "Meet the Inventor of the Electronic Spreadsheet," filmed at TEDxBeaconStreet, November 2016, TED Conferences, https://www.ted.com/talks/dan_bricklin_meet_the_inventor_of_the_electronic_spreadsheet/transcript.

42. Dan Bricklin, "The Idea," personal website, accessed April 9, 2020, http://www.bricklin.com/history/saiidea.htm.

43. Dan Bricklin, "Early Days," personal website, accessed April 9, 2020, http://www.bricklin.com/history/saiearly.htm.

44. Bricklin, "Meet the Inventor of the Electronic Spreadsheet."

45. Dan Bricklin, "The First Product," personal website, accessed April 9, 2020, http://www.bricklin.com/history/saiproduct1.htm.

46. Bricklin, "Meet the Inventor of the Electronic Spreadsheet."

47. Mitch Kapor, telephone interview with the authors, January 25, 2019.

48. William A. Sahlman, "Lotus Development Corp," Harvard Business School Case 285–094, January 1985, rev. February 1997, 6, https://www.hbs.edu/faculty/Pages/item.aspx?num=6195.

49. Ronald W. Coan, "Silicon Valley and Route 128: The Camelots of Economic Development," *Journal of Applied Research in Economic Development* (May 2013).

50. Susan Diesenhouse, "Real Estate: Lotus Development Is Completing a New Complex in the Town Where the Company Was Founded," *New York Times*, April 26, 1995, D23, https://www.nytimes.com/1995/04/26/business/real-estate-lotus-development-is-completing-a-new-complex-in-the.html.

51. Hiawatha Bray, "How Lotus Changed the Business World: Reunion Will Honor Innovator in Software and in the Workplace," *Boston Globe*, May 11, 2007.

52. Bray, "How Lotus Changed the Business World."

53. "Oral History of Mitch Kapor," catalog number 102657943, Computer History Museum, November 19, 2004, p. 18, accessed April 7, 2020, https://www.computerhistory.org/collections/catalog/102657943.

54. "Lotus Trims Workforce after Reorganization," C-Net, January 2, 2002, https://www.cnet.com/news/lotus-trims-work-force-after-reorganization.

55. Robin Chase, *Peers Inc.* (London: Headline Publishing Group, 2015), 8.

56. Arielle Duhaime-Ross, "Driven: How Zipcar's Founders Built and Lost a Car-Sharing Empire," *The Verge*, April 1, 2014, https://www.theverge.com/2014/4/1/5553910/driven-how-zipcars-founders-built-and-lost-a-car-sharing-empire.

57. Chase, *Peers Inc.*, 8–9.

58. Duhaime-Ross, "Driven."

59. Chase, *Peers Inc.*, 9.

60. Robin Chase, telephone interview with the authors, July 26, 2019.

61. Chase, *Peers Inc.*, 12.

62. Arun Sundararajan, telephone interview with the authors, April 25, 2019.

63. Kathryn Zickuhr, "Mobile Is the Needle; Social Is the Thread," Pew Research Center, October 18, 2012, https://www.pewresearch.org/internet/2012/10/18/mobile-is-the-needle-social-is-the-thread.

64. Glenn Urban, telephone interview with the authors, March 25, 2019.

65. Chase, *Peers Inc.*, 1.

66. "MITEI People: Antje Danielson," Massachusetts Institute of Technology Energy Institute, accessed February 28, 2020, http://energy.mit.edu/profile/antje-danielson.

67. Robin Chase, "About Robin," personal website, accessed February 28, 2020, http://www.robinchase.org/#about-robin.

68. Deborah Ancona and Cate Reavis, "Robin Chase, Zipcar and an Inconvenient Discovery," *MIT Sloan Management Review* (July 25, 2014): 13–14, https://www.thecasecentre.org/main/products/view?id=126468.

69. Jeremy Quittner, "Robin Chase: How I Survived a Huge Screw-up: When the Zipcar Founder Miscalculated Her Prices, There Was Only One Way to Salvage the Situation—Come Clean to Customers," Inc., February 28, 2013, https://www.inc.com/magazine/201303/how-i-got-started/robin-chase.html.

70. Duhaime-Ross, "Driven."

71. Agustino Fontevecchia, "Zipcar IPO: Raises $174M, Gains 67% to Lead US Markets," *Forbes*, April 14, 2011, https://www.forbes.com/sites/afontevecchia/2011/04/14/zipcar-ipo-raises-174m-gains-67-to-lead-us-markets/#451aff8d73dd.

72. Andrew Martin, "Car Sharing Catches On as Zipcar Sells to Avis," *New York Times*, January 2, 2013, https://dealbook.nytimes.com/2013/01/02/avis-to-buy-zipcar-for-500-million.

73. Duhaime-Ross, "Driven."

74. "Zipcar Locations," Zipcar.com, accessed November 15, 2020, https://www.zipcar.com/cities.

CHAPTER 7

1. Pierre Azoulay, telephone interview with the authors, April 4, 2018.

2. "The Kendall Story," Kendall Square Association, accessed March 7, 2020, www.kendallsq .org/history.

3. Justin Kaplan, "Baker Signs Bill Allotting Nearly Half a Billion to Mass. Life Science," WBUR. org, June 15, 2018, www.wbur.org/commonhealth/2018/06/15/baker-life-sciences-2.

4. "Mark Fishman," Harvard Stem Cell Institute, accessed March 7, 2020, https://hsci.harvard .edu/people/mark-fishman.

5. "*From Controversy to Cure*, Virtual Discussion, Coolidge Corner Theatre," August 25, 2020, YouTube, www.youtube.com/watch?v=ZG2NRCHCECA, accessed January 4, 2021.

6. Morton and Phyllis Keller, *Making Harvard Modern: The Rise of America's University* (New York: Oxford University Press, 2001), 234.

7. "*From Controversy to Cure*, Virtual Discussion, Coolidge Corner Theatre."

8. "History," Ragon Institute of Massachusetts General Hospital, Massachusetts Institute of Technology, and Harvard University, accessed March 7, 2020, http://www.ragoninstitute .org/about/history; "MIT and MGH Form Strategic Partnership to Address Major Challenges in Clinical Medicine," MIT News, October 15, 2014, https://news.mit.edu/2014 /mit-mgh-strategic-partnership-1015.

9. "Faculty Forum Online: *From Controversy to Cure*—Screening and Film Discussion," YouTube, November 3, 2020, https://www.youtube.com/watch?v=umZ9MrCM_io.

10. "Faculty Forum Online."

11. "Faculty Forum Online."

12. William Sahlman, telephone conversation with the authors, December 14, 2020.

13. Ali Khademhosseini, telephone interview with the authors, March 30, 2020.

14. Jonathan Shaw, "He Has Made the World a Safer Place," *Harvard Magazine*, June 1, 2018, https://www.harvardmagazine.com/2018/06/meselson-celebration.

15. Ariel Conn, "From DNA to Banning Biological Weapons with Matthew Meselson and Max Tegmark," part 1, podcast, Future of Life Institute, https://futureoflife.org/2019/02/28 /fli-podcast-part-1-from-dna-to-banning-biological-weapons-with-matthew-meselson-and -max-tegmark/?cn-reloaded=1&cn-reloaded=1.

16. Istvan Hargittai and Magdoina Hargittai, "Matthew Meselson," *Candid Science* 4 (2006): 45, Imperial College Press, https://doi.org/10.1142/9781860948855_0003.

17. Thomas Maniatis, "2004 Albert Lasker Special Achievement Award in Medical Science," Lasker Foundation, October 1, 2004, http://www.laskerfoundation.org/awards/show/public -policy-of-chemical-and-biological-weapons.

18. "Interview with Matthew Meselson," *BioEssays* 25, no. 12 (2003): 1239, accessed March 22, 2019, doi: 10.1002/bies.10374.

19. In recent years, researchers have begun manipulating mRNA to trigger the cell to make desired proteins. This approach is being used in vaccines against infectious diseases like COVID-19 and in treatments for conditions like cancer.

20. Matthew Cobb, "Who Discovered Messenger RNA?," *Current Biology Magazine* 25 (June 29, 2015): R531.

21. Conn, "From DNA to Banning Biological Weapons with Matthew Meselson and Max Tegmark."

22. Hargittai and Hargittai, "Matthew Meselson," 49–50.

23. Matthew Meselson, in-person interview with the authors, Cambridge, MA, April 24, 2019.

24. "Interview with Matthew Meselson," 1243.

25. Formally called the Convention on the Prohibition of the Development, Production, and Stockpiling of Bacteriological and Toxin Weapons and on Their Destruction (BTWC).

26. With notable exceptions being Iraq in 1988 and during the Syrian civil war, beginning in 2013. Julia Brooks et al., "Responding to Chemical Weapons Violations in Syria: Legal, Health, and Humanitarian Recommendations," *Conflict and Health* 12 (2018), https://www.ncbi.nlm.nih.gov/pmc/articles/PMC5817898.

27. Karen Weintraub, "Analysis Identifies Lethal Strain of Anthrax the Soviets Produced as Bioweapon," STAT, September 27, 2016, https://www.statnews.com/2016/09/27/anthrax-soviet-bioweapon.

28. John Durant, "Refrain from Using the Alphabet," *Becoming MIT* (Cambridge, MA: MIT Press, 2012), 151.

29. As quoted in Durant, "Refrain from Using the Alphabet," 151–153.

30. Sheldon Krimsky, *Genetic Alchemy: The Social History of the Recombinant DNA Controversy* (Cambridge, MA: MIT Press, 1982), 301.

31. Maryann Feldman and Nichola Lowe, "Consensus from Controversy: Cambridge's Biosafety Ordinance and the Anchoring of the Biotech Industry," *European Planning Studies* 16, no. 3 (2008): 399.

32. Krimsky, *Genetic Alchemy*, 305–307.

33. Feldman and Lowe, "Consensus from Controversy," 401.

34. Walter Gilbert, in-person interview with the authors, Cambridge, MA, February 13, 2019.

35. Feldman and Lowe, "Consensus from Controversy," 405.

36. James R. Hagerty, "Dutchman Who Had Trouble Finding Job at Home Became CEO of a U.S. Biotech Pioneer," *Wall Street Journal*, May 19, 2017.

37. "2019 Top 25 Employers," Community Development Department, City of Cambridge, https://www.cambridgema.gov/cdd/factsandmaps/economicdata/top25employers.

38. William Sahlman, telephone conversation with the authors, December 14, 2020.

39. "President Clinton Announcing the Completion of the First Survey of the Entire Human Genome," The White House at Work, June 26, 2000, https://clintonwhitehouse4.archives .gov/WH/Work/062600.html.

40. "Human Genome Project FAQ," National Human Genome Research Project, accessed August 7, 2019, https://www.genome.gov/human-genome-project/Completion-FAQ.

41. Seema Kumar, "Whitehead Scientists Enjoy Genome Sequence Milestone," MIT News, July 12, 2000, http://news.mit.edu/2000/whitehead-0712.

42. Eric Lander, in-person interview with the authors, Cambridge, MA, January 12, 2019.

43. David Page, in-person interview with the authors, Whitehead Institute for Biomedical Research, Cambridge, MA, June 21, 2019.

44. Walter Gilbert, in-person interview with the authors, Cambridge, MA, February 13, 2019.

45. Victor K. McElheny, *Drawing the Map of Life: Inside the Human Genome Project* (New York: Basic Books, 2012), 53–54.

46. Maynard V. Olson, "The Human Genome Project: A Player's Perspective," *Journal of Molecular Biology* 319 (2002): 931–942.

47. International Human Genome Sequencing Consortium, "Initial Sequencing and Analysis of the Human Genome," *Nature* 409 (February 15, 2001): 860–921, www.nature.com /articles/35057062.

48. Victor McElheny, *James Watson and the DNA Revolution* (Reading, MA: Perseus, 2003), 268–269.

49. International Human Genome Sequencing Consortium, "Finishing the Euchromatic Sequence of the Human Genome," *Nature* 431 (October 21, 2004): 931–945.

50. George Church, in-person interview with the authors, January 16, 2021.

51. Andrew Joseph, "New Vertex Cystic Fibrosis Drug Approved, Extending Treatments to 90% of Patients," STAT, October 21, 2019, https://www.statnews.com/2019/10/21/new -vertex-cystic-fibrosis-drug-approved-extending-treatments-to-90-of-patients.

52. McElheny, *Drawing the Map of Life*, 205.

53. Sarah Zhang, "How a Genealogy Website Led to the Alleged Golden State Killer," *The Atlantic*, April 27, 2018, https://www.theatlantic.com/science/archive/2018/04/golden-state -killer-east-area-rapist-dna-genealogy/559070.

54. "David Reich Lab: Ancient DNA, Biology, and Disease," Harvard University, accessed April 9, 2020, https://reich.hms.harvard.edu.

55. Karen Weintraub and Elizabeth Weise, "The Sprint to Create a COVID-19 Vaccine Started in January: The Finish Line Awaits," *USA Today*, September 11, 2020.

56. Robert Langer, email to the authors, April 11, 2020.

57. "Robert Langer, ScD," Department of Biological Engineering, School of Engineering, Massachusetts Institute of Technology, accessed February 15, 2020, https://be.mit.edu/directory/robert-langer.

58. "Scientists You Must Know: Robert Langer," video, Science History Institute, 2:17–3:00, https://www.sciencehistory.org/files/scientists-you-must-know-robert-langer.

59. "Scientists You Must Know: Robert Langer," 3:40–7:35

60. Robert S. Langer, public talk on his receipt of the 2019 Dreyfus Prize in the Chemical Sciences, Massachusetts Institute of Technology, September 26, 2019.

61. Luke Timmerman, "Life of a Scientific Entrepreneur: Bob Langer of MIT," The Long Run, Timmerman Report, March 26, 2019, 28:05, https://timmermanreport.com/2019/03/life-of-a-scientific-entrepreneur-bob-langer-of-mit-on-the-long-run.

62. Timmerman, "Life of a Scientific Entrepreneur: Bob Langer of MIT," 27:55–28:05.

63. Amanda Schaffer, "The Problem Solver," *MIT Technology Review* (April 21, 2015), https://www.technologyreview.com/s/536351/the-problem-solver.

64. Schaffer, "The Problem Solver."

65. Timmerman, "Life of a Scientific Entrepreneur: Bob Langer of MIT," 27:55–28:05.

66. "Avastin," FiercePharma, February 26, 2019, https://www.fiercepharma.com/special-report/4-avastin.

67. "Bevacizumab," National Cancer Institute, https://www.cancer.gov/about-cancer/treatment/drugs/bevacizumab.

68. Mark Guidera, "Guilford Gets OK to Market Cancer Wafer FDA's Green Light Called a Turning Point for Baltimore Firm; Brain Tumor Deterrent; Wall Street Applauds 'Big Achievement for a Small Company,'" *Baltimore Sun*, September 26, 1996.

69. "Important Questions about Gliadel Wafer," Arbor Pharmaceuticals, accessed February 15, 2020, https://gliadel.com/patient/questions-about-gliadel.php.

70. Charles A. Vacanti, "The History of Tissue Engineering," *Journal of Cellular and Molecular Medicine* 10, no. 3 (July 2006): 569–576, published online May 1, 2007.

71. Robert Langer and Jay Vacanti, "Tissue Engineering," *Science* 260, no. 5110 (May 14, 1993): 920–926, doi: 10.1126/science.8493529, https://science.sciencemag.org/content/260/5110/920.

72. Vacanti, "The History of Tissue Engineering," 569–576.

73. Ali Khademhosseini, telephone interview with the authors, March 30, 2020.

74. Rob Matheson, "Major Step for Implantable Drug-Delivery Device," MIT News, June 29, 2015, http://news.mit.edu/2015/implantable-drug-delivery-microchip-device-0629; "Daré Bioscience Closes Previously Announced Acquisition of Microchips Biotech with a First-in-Class Wireless, User-Controlled Drug Delivery Platform," Daré Bioscience, press release, November 21, 2019, https://www.globenewswire.com/news-release/2019/11/21/1950720/0

/en/Dar%C3%A9-Bioscience-Closes-Previously-Announced-Acquisition-of-Microchips
-Biotech-with-a-First-in-Class-Wireless-User-Controlled-Drug-Delivery-Platform.html.

75. Sharon Begley, "Meet One of the World's Most Groundbreaking Scientists. He's 34," STAT, November 6, 2015, https://www.statnews.com/2015/11/06/hollywood-inspired -scientist-rewrite-code-life.

76. Kevin Davies, *Editing Humanity: The CRISPR Revolution and the New Era of Genome Editing* (New York: Pegasus Books, 2020), 80–84.

77. Martin Jinek et al., "A Programmable Dual-RNA-Guided DNA Endonuclease in Adaptive Bacterial Immunity," *Science* 337, no. 6096 (August 17, 2012): 816–821, https://pubmed .ncbi.nlm.nih.gov/22745249.

78. Kevin Davies, *Editing Humanity: The CRISPR Revolution and the New Era of Genome Editing* (New York: Pegasus Books, 2020), 88.

79. Heidi Ledford and Ewen Callaway, "Pioneers of Revolutionary CRISPR Gene Editing Win Chemistry Nobel: Emmanuelle Charpentier and Jennifer Doudna Share the Award for Developing the Precise Genome-Editing Technology," *Nature* 586 (October 15, 2020): 346.

80. Haydar Frangoul et al., "CRISPR-Cas9 Gene Editing for Sickle Cell Disease and β-Thalassemia," *New England Journal of Medicine* (December 5, 2020), doi: 10.1056/NEJMoa2031054.

81. David Liu, telephone conversation with the authors, January 3, 2021.

82. Davies, *Editing Humanity*, 327.

83. Alexis Komor et al., "Programmable Editing of a Target Base in Genomic DNA without Double-Stranded DNA Cleavage," *Nature* 533 (April 20, 2016): 420–424.

84. Nicole Gaudelli et al., "Programmable Base Editing of A-T to G-C in Genomic DNA without DNA Cleavage," *Nature* 551 (November 23, 2017): 464–471.

85. Luke Koblan et al., "In Vivo Base Editing Rescues Hutchinson-Gilford Progeria Syndrome in Mice," *Nature* 589 (January 6, 2021): 608–614.

86. Andrew Anzalone et al., "Search-and-Replace Genome Editing without Double-Strand Breaks or Donor DNA," *Nature* 576 (October 21, 2019): 149–157.

87. Kevin Davies, telephone interview with the authors, January 18, 2021.

CHAPTER 8

1. Joan Baez, *And a Voice to Sing With: A Memoir* (New York: Simon & Schuster Paperbacks, 2009), 63.

2. Zach Goldhammer, "What Bob Dylan Learned in Harvard Square," The ARTery, WBGH, December 10, 2016, https://www.wbur.org/artery/2016/12/10/bob-dylan-harvard-square.

3. "May 9, 1974: Critic Declares Springsteen Future of Rock and Roll," MassMoments, Mass Humanities, accessed March 24, 2020, https://www.massmoments.org/moment-details /critic-declares-springsteen-future-of-rock-and-roll.html.

4. Earl Zuckerman, "This Date in History: First Football Game Was May 14, 1874," Channels: McGill University News and Events, McGill University, May 14, 2012, https://www.mcgill.ca/channels/news/date-history-first-football-game-was-may-14-1874-106694.

5. Parke H. Davis, *Football: The American Intercollegiate Game* (New York: Charles Scribner's Sons, 1911), 62–66, https://archive.org/details/footballamerican00davirich.

6. Ronald Smith, *Sports and Freedom: The Rise of Big-time College Athletics* (New York: Oxford University Press, 1988), 75.

7. Mark F. Bernstein, *Football: The Ivy League Origins of an American Obsession* (Philadelphia: University of Pennsylvania Press, 2001), 10. See also Zuckerman, "This Date in History: First Football Game Was May 14, 1874."

8. "The Foot-Ball Match," *Harvard Crimson*, May 22, 1874, https://www.thecrimson.com/article/1874/5/22/the-foot-ball-match-the-second-game.

9. Davis, *Football: The American Intercollegiate Game*, 64. See also Bernstein, *Football: The Ivy League Origins*, 10.

10. Zuckerman, "This Date in History: First Football Game Was May 14, 1874."

11. Smith, *Sports and Freedom*, 75.

12. Davis, *Football: The American Intercollegiate Game*, 62–66.

13. Smith, *Sports and Freedom*, 75.

14. Zuckerman, "This Date in History: First Football Game Was May 14, 1874."

15. Smith, *Sports and Freedom*, 76.

16. Paul Sweeney, "Tufts-Harvard First Game Story Breaks Nationally," Tufts Athletic Department, Tufts University, last modified 2004–2005, https://ase.tufts.edu/athletics/old/menFootball/press/2004-2005/firstgamebuzz.html.

17. Smith, *Sports and Freedom*, 76.

18. Bernstein, *Football: The Ivy League Origins*, 12.

19. "Birth of Pro Football," Pro Football Hall of Fame, https://www.profootballhof.com/football-history/birth-of-pro-football.

20. Bernstein, *Football: The Ivy League Origins*, 84.

21. Bernstein, *Football: The Ivy League Origins*, 35–38.

22. John Sayle Watterson, *College Football: History, Spectacle, Controversy* (Baltimore, MD: Johns Hopkins University Press, 2000), 23.

23. Brett Tomlinson, "Princeton's Role in the Birth of Thanksgiving Football," Princeton Alumni Weekly, November 27, 2014, https://paw.princeton.edu/article/throwbackthursday-princeton's-role-birth-thanksgiving-football.

24. Thomas G. Bergin, *The Game: The Harvard–Yale Football Rivalry, 1875–1983* (New Haven, CT: Yale University Press, 1984), 82.

25. Bernstein, *Football: The Ivy League Origins*, 84.

26. Christopher Kimball, in-person interview with the authors, Kimball's Milk Street office, Boston, MA, February 5, 2019.

27. Julia Child and Alex Prud'homme, *My Life in France* (New York: Anchor Books, 2006), 18.

28. Child and Prud'homme, *My Life in France*, 120.

29. "1963," Julia Child Foundation for Gastronomy and the Culinary Arts, accessed April 30, 2020, https://juliachildfoundation.org/1963-2.

30. Jacques Pépin, "Memories of a Friend, Sidekick and Foil," *New York Times*, August 15, 2012, https://www.nytimes.com/2012/08/15/dining/jacques-pepin-recalls-friendship-with-julia-child.html.

31. Dan Ackroyd, "The French Chef," video clip, *Saturday Night Live*, NBC, December 9, 1978, https://www.nbc.com/saturday-night-live/video/the-french-chef/n8667.

32. Kimball, in-person interview with the authors, February 5, 2019.

33. Barry Bluestone and Mary Huff Stevenson, *The Boston Renaissance: Race, Space, and Economic Change in an American Metropolis* (New York: Russell Sage Foundation, 2002), 1–3.

34. Sam Roberts, "Infamous 'Drop Dead' Was Never Said by Ford," *New York Times*, December 28, 2006, A30, https://www.nytimes.com/2006/12/28/nyregion/28veto.html.

35. Erik Lacitis, "'Turn Out the Lights': Message from 1971 Seattle Billboard Echoed in Head-Tax Debate," *Seattle Times*, May 21, 2018, https://www.seattletimes.com/seattle-news/turn-out-the-lights-message-from-1971-seattle-billboard-echoed-in-head-tax-debate.

36. Jane Holtz Kay, "Rouse-ification of Lower Manhattan: South Street Seaport a Mix of Excellent and Not-So-Good Design," *Christian Science Monitor*, September 16, 1983, 9.

37. Alex Krieger, "The Thompson Sampler," *Architecture Boston* 14, no. 1 (Spring 2011): 20.

38. Robert Campbell, *AIA Journal* (June 1981): 28.

39. Walter Gropius, Walter, Jean B. Fletcher, Norman C. Fletcher, John C. Harkness, Sarah P. Harkness, Louis A. McMillen, and Benjamin Thompson, eds., *The Architect's Collaborative, 1945 to 1965* (New York: Architectural Book Publishing Co., 1966), 37–41.

40. Anthony Thompson, "The Thompson Sampler," *Architecture Boston* 14, no. 1 (Spring 2011): 17.

41. "About Marimekko," Marimekko.com, accessed March 27, 2020, https://company.marimekko.com/en/about-marimekko/history.

42. David Fixler, "The Pied Piper of Modernism: The Man, the Myths, the Movement," *Architecture Boston* 16, no. 2 (Summer 2013): 31–33.

43. "Shopping Around in Harvard Square," *Interiors* (May 1970): 109.

44. "Harvest: 40 Years in Harvard Square," Harvest Restaurant, accessed March 27, 2020, https://harvestcambridge.com/about-harvest/anniversary.

45. "Benjamin C. Thompson (1918–2002)," Mount Auburn Cemetery, August 2, 2011, https://mountauburn.org/benjamin-c-thompson-1918-2002.

46. David W. Dunlap, "Benjamin C. Thompson, 84, Architect of Festive Urban Marketplaces, Is Dead," *New York Times*, August 20, 2002.

47. "Domestic Grosses Adjusted for Ticket Price Inflation," Box Office Mojo, accessed February 9, 2019, https://www.boxofficemojo.com/alltime/adjusted.htm.

48. "America's 100 Greatest Love Stories or Romantic Films," American Film Institute, accessed February 9, 2019, https://www.filmsite.org/afi100loves1.html.

49. Ty Burr, in-person interview with the authors, Johnny's Luncheonette, Newton, MA, February 22, 2019.

50. Vincent Canby, "Perfection and a *Love Story*: Erich Segal's Romantic Tale Begins Run," *New York Times*, December 18, 1970.

51. Pamela Jaffee, Personal correspondence, Harper Collins, July 8, 2019, confirming early February publication and release date (although Wikipedia and other sources claim a February 14 publication date).

52. Margalit Fox, "Erich Segal, *Love Story* Author, Dies at 72," *New York Times*, January 19, 2010.

53. "Media Relations Policies," Harvard University, accessed February 22, 2019, https://www.harvard.edu/media-relations/policies.

54. Sarah Sweeney, "O'Neal, MacGraw Revisit Youthful *Love*," *Harvard Gazette*, February 2, 2016, https://news.harvard.edu/gazette/story/2016/02/oneal-macgraw-revisit-youthful-love.

55. Philip Marcelo, "*Love Story* Actors Return to Harvard 45 Years Later," *Seattle Times*, January 31, 2016, https://www.seattletimes.com/nation-world/love-story-actors-return-to-harvard-45-years-later.

56. "Brazelton: Listening to Children—and Their Parents," National Public Radio, May 10, 2007, https://www.npr.org/templates/story/story.php?storyId=10098366.

57. Heidelise Als, in-person interview with the authors, Boston Children's Hospital, Boston, MA, August 23, 2019.

58. Sandra Blakeslee, "Dr. T. Berry Brazelton, Who Explored Babies' Mental Growth, Dies at 99," *New York Times*, March 14, 2018.

59. Joshua Sparrow, teleconference with the authors, April 4, 2019.

60. T. Berry Brazelton, *Learning to Listen: A Life Caring for Children* (Boston: Da Capo Press, 2013), 6.

61. Sparrow, teleconference with the authors, April 4, 2019.

62. Heidelise Als, in-person interview with the authors, Boston Children's Hospital, Boston, MA, August 23, 2019.

63. "*Car Talk*'s Tom: A Biography," WBUR News, November 03, 2014.

64. Bryan Marquard, "*Car Talk* Co-host Tom Magliozzi Dies at 77," *Boston Globe*, November 3, 2014.

65. "The History of *Car Talk*," Cartalk.com, accessed November 17, 2018, https://www .cartalk.com/content/history-car-talk.

66. "*Car Talk*'s Tom: A Biography."

67. "*Car Talk*," Peabody Awards, accessed March 27, 2020, http://www.peabodyawards.com /award-profile/car-talk.

68. "Ray Magliozzi," Automotive Hall of Fame, accessed March 27, 2020, http://www .automotivehalloffame.org/honoree/ray-magliozzi; "Tom Magliozzi," Automotive Hall of Fame, accessed March 27 2020, http://www.automotivehalloffame.org/honoree/tom-magliozzi.

69. Doug Berman, telephone interview with the authors, December 31, 2019.

70. "*Car Talk*'s Tom: A Biography."

71. Noam Cohen, "Tom Magliozzi, One Half of the Jovial Brothers on *Car Talk*, Dies at 77," *New York Times*, November 3, 2014.

72. Henry Louis Gates Jr., *Faces of America: How 12 Extraordinary People Discovered Their Pasts* (New York: NYU Press, 2010), 92, https://muse.jhu.edu/book/12513.

73. Charlotte Smith, "The Great Connector," *The Strad* (January 2018): 34, www.thestrad .com.

74. Gates, *Faces of America*, 97.

75. Janet Tassel, "Yo-Yo Ma's Journeys," *Harvard Magazine*, March 2000, https://harvard magazine.com/2000/03/yo-yo-mas-journeys-html.

76. Benjamin C. Burns, "Yo-Yo Ma Goes beyond the Music," *Harvard Crimson*, February 12, 2009, https://www.thecrimson.com/article/2009/2/12/yo-yo-ma-goes-beyond-the-music.

77. Janet Tassel, "Yo-Yo Ma's Journeys," *Harvard Magazine*, March 2000, https://harvard magazine.com/2000/03/yo-yo-mas-journeys-html.

78. "SilkRoad: About Us," Silkroad, accessed February 18, 2019, https://www.silkroad.org /about.

79. Alex Ross, "Yo-Yo Ma's Days of Action," *New Yorker*, December 17, 2018.

80. Peter Dizikes, "Yo-Yo Ma Calls for 'Culture in Action' to Build a Better World," MIT News, March 20, 2018, http://news.mit.edu/2018/compton-lecture-yo-yo-ma-culture-action-build.

CHAPTER 9

1. Edward Glaeser, *Triumph of the City* (New York: Penguin Books, 2011), 27–29.

2. "City of Cambridge, Massachusetts Statistical Profile," Cambridge Community Development Department, June 1, 2011, 57, www.cambridgema.gov/~/media/Files/CDD/Factsand Maps/profiles/demo_profile_statistical.pdf.

3. Glaeser, *Triumph of the City*, 235.

4. Glaeser, *Triumph of the City*, 232.

5. "2019 Top 25 Employers," City of Cambridge, MA, https://www.cambridgema.gov /CDD/factsandmaps/economicdata/top25employers.

6. Barry Bluestone, telephone interview with the authors, March 29, 2020.

7. "Population Density for U.S. Cities Statistics," *Governing Magazine*, accessed July 26, 2019, www.governing.com/gov-data/population-density-land-area-cities-map.html.

8. Robin Chase, telephone interview with the authors, July 26, 2019.

9. Glaeser, *Triumph of the City*, 235.

10. Paolo Malanima, "The Italian Renaissance Economy (1250–1600)," paper delivered at the International Conference at Villa la Pietra, Florence, May 10–12, 2008, 3–4, http://citeseerx .ist.psu.edu/viewdoc/download?doi=10.1.1.489.362&rep=rep1&type=pdf.

11. "City of Cambridge, Massachusetts Statistical Profile," Cambridge Community Development Department, June 1, 2011, 17.

12. Jason Lange, Yeganeh Torbati, "U.S. Foreign-Born Population Swells to Highest in over a Century," Reuters, September 13, 2018, https://www.reuters.com/article/us-usa-immigra tion-data/u-s-foreign-born-population-swells-to-highest-in-over-a-century-idUSKCN1L T2HZ.

13. "Public Schools Language Spoken at Home 2014–2019," Cambridge Community Development Department, accessed July 30, 2019, https://data.cambridgema.gov/Planning /Public-Schools-Language-Spoken-at-Home-2014-2019/vkjy-igvt.

14. Richard Florida, *The New Urban Crisis: How Our Cities Are Increasing Inequality, Deepening Segregation, and Failing the Middle Class—and What We Can Do about It* (New York: Basic Books, 2017), xv; see also Richard Florida, *The Rise of the Creative Class: And How It's Transforming Work, Leisure, Community, and Everyday Life* (New York: Basic Books, 2002).

15. Colleen Walsh, "Reviving the Philosophy Chamber," *Harvard Gazette*, May 16, 2017, https://news.harvard.edu/gazette/story/2017/05/reviving-the-philosophy-chamber.

16. S. William Green, "Lawrence Scientific School Marked Era in U.S. Intellectual History," *Harvard Crimson*, February 21, 1948, https://www.thecrimson.com/article/1948/2/21 /lawrence-scientific-school-marked-era-in.

17. The gift amounted to $16 million by 1949, when Harvard was able to take full possession of the principal, and it has supported engineering and applied science professorships ever since. "Gordon McKay," School of Engineering and Applied Sciences, Harvard University, accessed March 27, 2020, https://www.seas.harvard.edu/about-seas/history-seas/historical-profiles /gordon-mckay.

18. Karl Zinsmeister, "George Eastman," Philanthropy Roundtable, accessed July 26, 2019, https://www.philanthropyroundtable.org/almanac/people/hall-of-fame/detail/george-eastman.

19. Katherine Igoe, "These 15 Startups Raised More Than $50M from Out-of-State VCs," Bost-Inno, December 18, 2018, American City Business Journals, https://www.americaninno.com/boston/end-of-year-boston/these-15-startups-raised-more-than-50m-from-out-of-state-vcs.

20. Alan Mallach and Lavea Brachman, *Regenerating America's Legacy Cities* (Cambridge, MA: Lincoln Institute of Land Policy, 2013), 31.

21. Mallach and Brachman, *Regenerating America's Legacy Cities*.

22. The Museum of Science Bridge, the Longfellow Bridge, the Massachusetts Avenue Bridge, the BU (Boston University) Bridge, the River Street and Western Avenue bridges, the Weeks Footbridge, the Anderson Memorial Bridge connecting North Harvard and JFK streets, and the Eliot Bridge from Soldiers Field Road.

23. Barry Bluestone, telephone interview with the authors, March 29, 2020.

24. Cambridge Election Commission, accessed December 30, 2020, https://www.cambridgema.gov/-/media/Files/electioncommission/2020statepresidential/november2020turnout.pdf; James M. Lindsay, "The 2020 Election by the Numbers," blog post, Council on Foreign Relations, December 15, 2020, https://www.cfr.org/blog/2020-election-numbers; "Voter Turnout Statistics," Secretary of the Commonwealth of Massachusetts, accessed December 30, 2020, https://www.sec.state.ma.us/ele/elevoterturnoutstats/voterturnoutstats.htm.

25. Kevin Roose, "Silicon Valley Is Over, Says Silicon Valley," *New York Times*, March 4, 2018, https://www.nytimes.com/2018/03/04/technology/silicon-valley-midwest.html; Sunil Rajaraman, "It's Time to Leave San Francisco: You Are Old School, and You Long for 2012," personal blog, March 12, 2018, https://thebolditalic.com/its-time-to-leave-san-francisco-2a5a74f42433; Azat Mardan, "Don't Move to San Francisco or 7 Reasons Why You Should Avoid Living in the Bay Area," personal blog, https://azat.co/blog/not-sf.

26. Greg McCarriston, "Affordable Inland Metros Drew People from San Francisco, New York and Los Angeles," Redfin, February 7, 2018, https://www.redfin.com/blog/q4-migration-report.

27. "Cambridge Housing Profile," Cambridge Community Development Department, 2016, 7.

28. Jenna Russell and Meghan E. Irons, "Cambridge High School Struggles with Equal Access to AP Classes," *Boston Globe*, February 20, 2020, https://www.bostonglobe.com/2020/02/20/metro/cambridge-high-school-struggles-with-equal-access-ap-classes/https://www.bostonglobe.com/2020/02/20/metro/cambridge-high-school-struggles-with-equal-access-ap-classes.

29. Sharon Begley, "Three Star Scientists Announce Plan to Solve Biotech's 'Missing Women' Problem," STAT, January 29, 2020, https://www.statnews.com/2020/01/29/biotechs-missing-women.

30. "Ensuring Scientific Justice by Building Bridges to Minority Communities Is Centerpiece of BIOEquality Agenda," press release, August 6, 2020, https://www.bio.org/press-release/ensuring-scientific-justice-building-bridges-minority-communities-centerpiece.

31. Florida, *The New Urban Crisis*, xviii.

32. Florida, *The New Urban Crisis*, 10–11.

33. "Per Pupil Expenditures, All Funds," Massachusetts Department of Elementary and Secondary Education, School and District Profiles, accessed August 17, 2019, http://profiles.doe.mass.edu/statereport/ppx.aspx.

34. "Community Preservation Act," City of Cambridge, MA, accessed August 17, 2019, https://www.cambridgema.gov/Departments/communitypreservationact.

35. Hannah Harn, "Deadline Approaches for Mayor's Summer Youth Employment Program," *Cambridge Chronicle*, May 1, 2018, https://cambridge.wickedlocal.com/news/20180501/deadline-approaches-for-mayors-summer-youth-employment-program.

36. Bluestone, telephone interview with the authors, March 29, 2020.

INDEX